DESIGN AND CONSTRUCTION OF SOIL ANCHOR PLATES

DESIGN AND CONSTRUCTION OF SOIL ANCHOR PLATES

HAMED NIROUMAND

KHAIRUL ANUAR KASSIM

ELSEVIER

AMSTERDAM • BOSTON • HEIDELBERG • LONDON
NEW YORK • OXFORD • PARIS • SAN DIEGO
SAN FRANCISCO • SINGAPORE • SYDNEY • TOKYO
Butterworth-Heinemann is an imprint of Elsevier

Butterworth-Heinemann is an imprint of Elsevier
The Boulevard, Langford Lane, Kidlington, Oxford OX5 1GB, United Kingdom
50 Hampshire Street, 5th Floor, Cambridge, MA 02139, United States

British Library Cataloguing-in-Publication Data
A catalogue record for this book is available from the British Library.

Library of Congress Cataloging-in-Publication Data
A catalog record for this book is available from the Library of Congress.

ISBN: 978-0-12-420115-6

For Information on all Butterworth-Heinemann publications
visit our website at https://www.elsevier.com/

Working together
to grow libraries in
developing countries

www.elsevier.com • www.bookaid.org

Publisher: Joe Hayton
Acquisition Editor: Kenneth P. McCombs
Editorial Project Manager: Peter Jardim
Production Project Manager: Kiruthika Govindaraju
Designer: Mark Rogers

Typeset by MPS Limited, Chennai, India

CONTENTS

ABOUT THE AUTHORS

Dr Hamed Niroumand is an assistant professor at the Department of Civil Engineering, Buein Zahra Technical University. He is currently the Vice Chancellor for the Research and Academic section of Buein Zahra Technical University. His main fields of research are geotechnical engineering, earth anchors, deep foundation, numerical analysis, sustainable development, and nanomaterials. He has  been a project manager and professional engineer in various geotechnical and earth building projects. Between 2011 and 2015 he received four medals and international awards for his inventions and research; first place for research at the national Iranian young inventor and researcher festival 2012 and the first place for research at the national Iranian youth festival in 2012 and 2013. In 2016 he received the best researcher award from the Ministry of Road and Urban Development in 2016. He has been the chairman and head director of the international/national conferences of civil engineering for close to 20 cases held in various countries. Dr Niroumand has chaired sessions in several international/national conferences and festivals in various countries and presented various research papers in many conferences around the world. He has published around 200 papers in journals and conferences. He is also an editorial team and reviewer in scientific journals. He has invented around 15 inventions that are patent/patent pending at this moment.

Personal website: www.niroomand.net

Dr Khairul Anuar Kassim is a Professor at the Faculty of Civil Engineering, Universiti Teknologi Malaysia (UTM). He is currently the Dean for the Faculty of Civil Engineering at UTM. Dr Khairul Anuar is an active researcher, securing various research grants amounting to more than RM 1 million. He has produced more than 200 papers, including conference papers, in indexed and  nonindexed journals. Being a research-oriented person, he has received awards from national and international bodies. He is also a member of several national and international organizations and professional bodies.

PREFACE

Anchor plates are geotechnical devices that have been constructed to support geotechnical structures. They are devised to prevent the overturning of structures experiencing lateral, inclined, and uplift loads. Anchor plates can be applied in a wide variety of uses, such as stabilization systems in foundations, retaining walls, sea walls, and pipe lines. In this book, we have focused on the structural design of anchor plates, their uplift and bearing capacity, as the anchors are primarily designed to resist outwardly-directed loads imposed onto the foundation(s) of structures.

This book consists of three main sections. The first section gives an introduction to all types of earth anchors. The second section gives calculation equations and outlines the requirements of anchor plates for use in sandy soils, and plates embedded in clays. The third section explains all the features of anchor plates that can be used in multilayer soils. Throughout the text, we compare and contrast some of the current theories and modeling for calculating the capacity of anchor plates. Alongside the textual elements, this book also contains several examples and photos to help illustrate the details of anchor plates and their usage. We have aimed to present the information in a user-friendly manner that is easy to follow and practice.

The book contains nine chapters. Chapters "Anchors" and "Anchor Plates" describes and review all the types of earth anchors that have geotechnical applications. Chapters "Horizontal Anchor Plates in Cohesionless Soil" and "Horizontal Anchor Plates in Cohesive Soil" investigate horizontal anchor plates as used in various types of soils such as clays and sands—both their features and their capacity. Chapters "Vertical Anchor Plates in Cohesionless Soil" and "Vertical Anchor Plates in Cohesive Soil" present some useful information regarding vertical anchor plates. These chapters describe the various types of vertical anchor plates, their capacity, and their common applications. Chapters "Inclined Anchor Plates in Cohesionless Soil" and "Inclined Anchor Plates in Cohesive Soil" describe inclined anchor plates and their usage in different types of soil such as clays and sands. These chapters describe the installation process and detail the capacity of inclined anchor plates, including examples where relevant. Chapter "Anchor Plates in Multilayer Soil" describes the use of anchor plates in multilayer soils.

The authors of this book hope that they have succeeded in providing readers with useful information about different types of anchor plates, given that they have a wide variety of application in crucial construction projects such as retaining walls, sea walls, and foundations of structures. Thus, the authors hope that this book fosters the further development of anchor plates and optimization of their applications.

Hamed Niroumand

CHAPTER 1

Anchors

1.1 INTRODUCTION

The choice of foundation systems has an important role in the design of many structures, to ensure they support any vertical or horizontal loads. Structures such as seawalls, transmission towers, tunnels, buried pipelines, and retaining walls are subjected to pullout forces and overturning moments. Thus, in the building process, the development of guidelines for the design and installation of anchor systems is one way to improve the performance of foundation systems.

For structures such as we mentioned above, using tension members can prove to be an economic design solution. Tension members are a type of soil anchor, and can also be used for tieback resistance in waterfront structures, pressure soil structures, and also against thermal stresses. Tension members should be fixed to the structure and then embedded into the ground to a considerable depth in order to resist uplifting forces.

In general, soil anchors are foundation systems used to transmit forces from the structure to the ground, in order to resist overturning moments and pullout forces which can threaten a structure's stability. The shear strength and dead weight of the soil surrounding an anchor are factors that improve an anchor's strength. Traditionally, soil anchors were divided into only three categories: grouted anchors, helical anchors, and anchor plates; however, a new self-driven anchor placed into the ground without excavation and grout has also been recently added to the list. According to the method of load transfer from the anchor to the surrounding soil, types of soil anchors in geotechnical engineering can be categorized as

- Grouted anchors
- Helical anchors
- Anchor plates
- Anchor piles
- Irregular shape anchors, used as self-driven anchors.

Design and Construction of Soil Anchor Plates.
DOI: http://dx.doi.org/10.1016/B978-0-12-420115-6.00001-1
1

1.2 GROUTED ANCHORS

A grouted soil anchor is installed in grout-filled drilled holes in the soil or rock and transmits an applied tensile load into the soil or rock. Grouted anchors are structural elements where a grout body is installed in the subsoil or rock by injecting grouting mortar around the rear part of the steel tendon. The grouted body is connected to the structure, or the rock section to be anchored, using steel or fiber tendon(s) and the anchor head. Any load to be taken up by the grouted anchor is passed into the subsoil, instead of over the entire length of the anchor, but only in the area of the grouted body. The steel tendon section, where the soil anchor is free to expand, is the free tendon length. This section acts like a spring that can be pretensioned to the structure against the subsoil.

Grouted anchors are subjected to tension only and their load capacity is checked by tensioning. Bonded deep into the ground using cementatious grout, these soil anchors transfer the necessary forces to restrain walls from overturning, water tanks and towers from uplifting forces, dams from rotating and other naturally- or phenomenally occurring forces applied to structures. The grouted anchor capacity is a function of the steel capacity as well as the geotechnical holding capacity. The steel capacity should be limited to 80% maximum test load and 60% lock-off load for permanent applications. However, the geotechnical holding capacity is a function of the ground bond stress characteristics that can be optimized by field procedures. Mainly, the cementation of a grouted body ensures a radial tie-in of the anchor tendon in the soil and also provides simple corrosion protection.

A grouted anchor consists of two zones:
- Anchor bond zone: The length along which the grout adheres to the soil and results in load transfer that can be placed above the tendon bond zone.
- Tendon bond zone: The length along which the bar/tendons are bonded to the grout and transfer the load.

  The basic components of a grouted anchor are:
- Anchorage
- Free stressing length
- Bond length.

The anchorage consists of the anchor head, bearing plate, and the trumpet that transmits the prestressing force from the prestressing steel to the soil or the structure (Figs. 1.1 and 1.2).

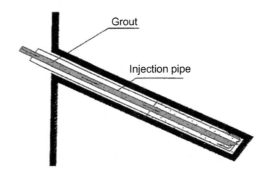

Figure 1.1 A grouted anchor.

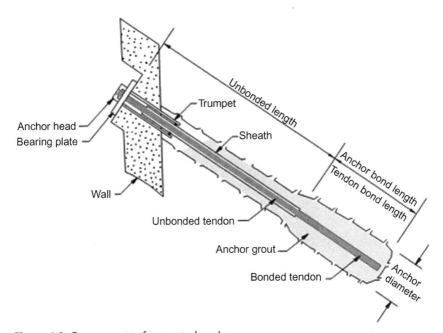

Figure 1.2 Components of a grouted anchor.

Four types of grouted anchors (Fig. 1.3) commonly used by engineers are
- Straight shaft gravity grouted anchors
- Straight shaft pressure grouted anchors
- Postgrouted anchors
- Under reamed anchors.

Each of the four types of grouted anchor is best suited to a specific purpose. For example, straight shaft pressure grouted anchors are used for coarse granular soils or weak and weathered rocks. In postgrouted

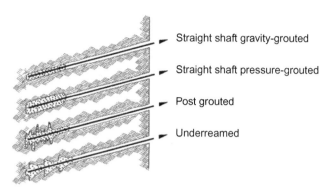

Figure 1.3 Four types of grouted anchors.

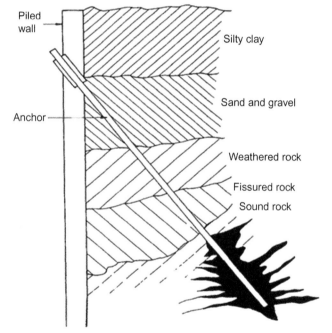

Figure 1.4 Grouted anchors used in sea walls.

anchors, in order to enlarge the body of the straight shaft gravity grouted anchors, multiple grout injections are used. On the other hand, the capacity of grout anchors under compression or tension is based on the bearing capacity of the soil or the structure in which they are installed.

Grouted anchors can be used in different geotechnical projects such as seawalls (Fig. 1.4), retaining walls (Fig. 1.5) and the foundation of pipelines and transmission towers to aid resistance against overturning.

Figure 1.5 Grouted anchors used in retaining walls.

1.3 HELICAL ANCHORS

Helical anchors have been used in the earth-boring industry for more than 170 years and bring "new" solutions to the soil stabilization and foundation industry. Sporadic use of helical anchors has been documented throughout the 19th and early 20th centuries, mainly for supporting structures and bridges built upon weak or wet soil. When hydraulic motors became readily available in the 1960s, allowing for easy and fast installation of helical anchors, their popularity flourished. Electric utility companies began to use helical anchors for tie down anchors on transmission towers and for guy wires on utility poles. Helical anchors are ideal for applications where there is a need to resist both tension and axial compression forces. Some examples of structures with a combination of these forces are metal buildings, canopies, and monopole telecommunication tower foundations. Current uses of helical anchors include underpinning foundations for commercial and residential structures, foundation repair, light standards, retaining walls, tieback anchors, pipeline and pumping equipment supports, elevated walkways and bridge abutments, along with numerous uses in the electric utility industry (Niroumand et al., 2013). Often helical anchors are the best solution for a foundation repair project because of one or more of the following factors:

- Ease of installation
- Little to no vibration
- Immediate load transfer upon installation

- Installed torque correlates to capacity
- Load easily tested to verify capacity
- Installs below active soils
- Can be installed in all weather conditions
- Little to no disturbance to the job site.

Helical anchors are used to resist tensile loads. Under downward-loading conditions, helical anchors can supply additional bearing capacity to the foundation. Helical anchors are screw forms combined with steel shafts and a series of helical steel plates that are attached to a pitch and are screwed into the soil until there is enough torque resistance to support the tieback wall systems and soil nail wall systems. Screw anchors are used for resistance in front of vertically uplifting loads in soils (Figs. 1.6–1.8). They can be installed in inclined and vertical positions. To reduce the disturbance in the soil, multiple helical anchors can be used, in which the upper helical anchors follow the lower ones.

A helical anchor is a deep, segmented foundation system with helical bearing plates welded to a central steel shaft. The load is transferred from the shaft to the soil through these bearing plates. Helical plates are spaced at distances far enough apart so that they can function independently as individual bearing elements. Consequently, the capacity of a particular helix

Figure 1.6 A single helix anchor.

Figure 1.7 A single helical anchor supporting a retaining wall.

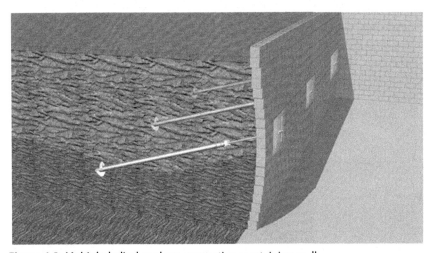

Figure 1.8 Multiple helical anchors supporting a retaining wall.

on a helical anchor shaft is not influenced by the helix above or below it. The capacity of a helical anchor under compression or tension is directly dependent on the bearing capacity of the helical plates against the soil in which it is located, hence the capacity of helical anchors increases as the area of the helical plates and the ground friction angle increases. This system helps the grout gain capacity and obviates disposal of soils, which becomes more important in contaminated soils. Figs. 1.9 and 1.10 show images of helical anchors. The installation of helical anchors is shown in Fig. 1.6.

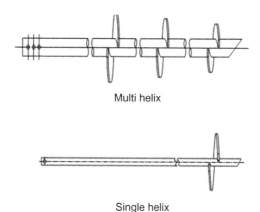

Multi helix

Single helix

Figure 1.9 Types of helical anchors.

Figure 1.10 Multiple helical anchors.

1.4 ANCHOR PLATES

For transmission towers, masts, and structures subjected to buoyancy effects, anchor plates are the most useful of soil anchors. Anchor plates consist of light structural elements that are used to resist against uplift forces. The way anchor plates are installed depends on the direction of the load applied. For instance, in resistance against vertical uplifting

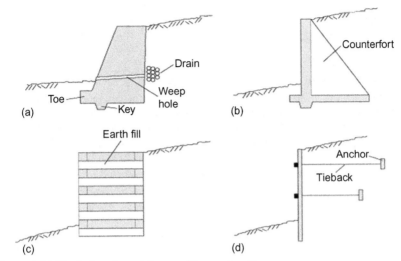

Figure 1.11 Vertical anchor plates used in construction.

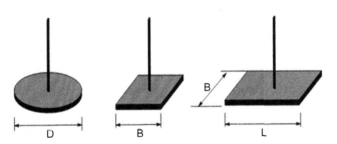

Figure 1.12 Different shapes of anchor plates.

loads, horizontal anchor plates are embedded; whereas vertical anchor plates are used to resist horizontal uplifting loads. Inclined axial loads, on the other hand, are used to resist axial pullout loads. For installation, the soil should be excavated to the required depth and then backfilled with the ground soil after placing the anchor plate (Fig. 1.11). Anchor plates are installed in excavated trenches as support for retaining structures. The shape of the anchor plate, such as square, strip or circular, is determined based on the bearing capacity of the anchor plate and the inflicted tension against the soil in which it is located (Fig. 1.12). These anchor plates are attached to tie rods that may either be driven or placed through augured holes (Fig. 1.13).

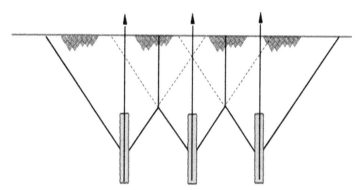

Figure 1.13 Interaction of cone in anchor plates.

1.5 ANCHOR PILES

Anchor piles are a type of soil anchor that can be used in various applications subjected to tension or compression loads. Piles (Fig. 1.14) are used to uplift problematic and downward loads. Anchor piles can be used to transmit an axial load to deeper soils, resist lateral loading, and resist uplift loading in practice. Piles can be defined into three types: drilled shaft, auger cast piles, and driven piles. Driven piles are defined into various types such as wood piles, steel piles, concrete piles, and composite piles.

1.6 IRREGULAR SHAPE ANCHOR AS A SELF-DRIVEN ANCHOR

The irregular shape anchor (Fig. 1.15) is an innovative, state-of-the-art self-driven anchor with an anchor plate that can be installed in the soil without any need for grout and excavation. The irregular shape anchor increases the speed of construction and is very cost effective. It is installed vertically by driving with a rod to the desired depth, then the rod is withdrawn and the cable is tensioned to rotate the irregular shape anchor through an angle of 90° into its final position Niroumand and Kassim, 2013 and Niroumand and Kassim, 2014.

1.7 SOIL ANCHOR APPLICATIONS

In construction foundations subjected to uplift, piles and drilled shafts can be used to support the downward load and resist against uplift forces.

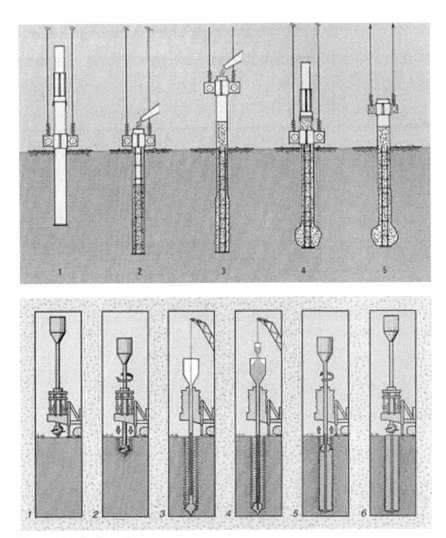

Figure 1.14 Anchor pile.

Although anchored systems are usually used for temporary applications, their service life is estimated to be about 7590 years, however corrosion protection methods are necessary. Anchor systems are helpful in earth structures such as transmission towers (Fig. 1.16), pipelines (Fig. 1.17), foundations (Fig. 1.18), sea walls (Fig. 1.19), retaining walls (Fig. 1.20), and other soil structures.

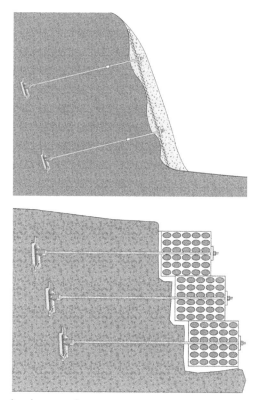

Figure 1.15 Irregular shape anchor.

Figure 1.16 Helical anchors used in transmission towers.

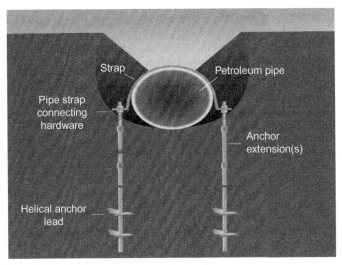

Figure 1.17 Multiple helical anchors used in pipelines.

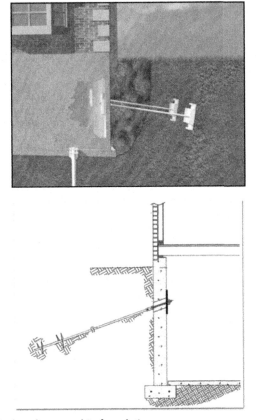

Figure 1.18 Helical anchors used in foundations.

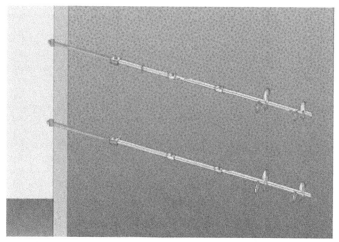

Figure 1.19 Helical anchors used in a seawall.

Figure 1.20 Helical anchors used in a retaining wall.

1.8 TIE DOWN STRUCTURES

In resisting against vertical uplift forces caused by hydrostatic or overturning loads, anchored systems can be useful to provide additional resistance to overturning, sliding and earthquake loading.

1.9 CONCLUSION

Grouted anchors need to be grouted during construction, which is one of their limitations, particularly in cold climate zones because the weather inhibits installation of the grouting. This system operates at a low speed because of the necessary duration required to sufficiently install the grout and also due to excavation requirements. Although many anchor plates do not require grouting, they also have limitations such as excavation requirements, low speed in construction, and their applications are few due to insufficient anchoring to counteract the uplift forces. Helical anchors are a good system because they do not need grout and excavation and have a satisfactory speed, but their applications are also limited. The irregular shape anchor is a new self-driven anchor that can be installed into soil without the need to grout or excavate; however, it can be used in tension structures only.

REFERENCES

Das, B.M., 1990. Earth Anchors. Elsevier, Amsterdam.

Niroumand, H., Kassim, K.A., 2013. Pullout capacity of irregular shape anchor in sand. Measurement 46 (10), 3876—3882.

Niroumand, H., Kassim, K.A., 2014. Uplift of irregular shape anchor in cohesion less soil. Arabian J. Sci. Eng. 39 (5), 3511—3524.

Niroumand, et al., 2013. Systematic review of screw anchors in cohesionless soils. Soil Mech. Found. Eng. 50 (5), 212—217.

CHAPTER 2

Anchor Plates

2.1 INTRODUCTION

For more than 50 years, various theories have been developed to analyze and predict the performance of soil anchor plates in different types of soils. Anchor plates are a type of soil anchor made of various materials such as steel plates, timber sheets, fiber reinforced polymer and precast concrete slabs. They are used to resist vertical, horizontal, and inclined loads in various geotechnical projects. Fitted to endure horizontal loads, this type of soil anchor is installed in various structures such as retaining walls, sea walls and related projects. Other types of anchor plates are used to resist uplift loads in various projects such as transmission towers, pipelines, and related structures. Based on these applications, inclined anchor plates are very important in various geotechnical projects, such as transmission towers and tents. Anchor plates can be used only in the soils of various geotechnical projects because they need to excavate the soil to install the anchor and then backfill it in, therefore they are termed "soil anchor plates." The new types of soil anchor plates use geosynthetics and soil reinforcement materials, and they have improved the quality and performance of soil anchor plates in various geotechnical projects.

2.2 TYPES OF ANCHOR PLATES

Published research on this topic is limited, but those that do exist have categorized soil anchor plates based on two aspects: (1) shape factor, and (2) applications. The shape of soil anchor plates can again be separated into four types as shown in Fig. 2.1. The shape types are rectangular, strip, circular and square plates. Subgroups of the application of soil anchor plate are horizontal, inclined and vertical anchor plates, as illustrated in Fig. 2.2.

2.2.1 Types of Anchor Plates Based on Applications
2.2.1.1 Horizontal Anchor Plate
The horizontal anchor plate may be used in the construction of various geotechnical projects subjected to uplift and pullout loads in cohesive or

Design and Construction of Soil Anchor Plates.
DOI: http://dx.doi.org/10.1016/B978-0-12-420115-6.00002-3

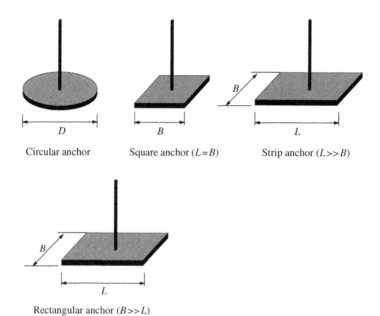

Circular anchor   Square anchor ($L=B$)   Strip anchor ($L>>B$)

Rectangular anchor ($B>>L$)

Figure 2.1 Soil anchor plates based on various shapes by Merifield and Sloan (2006).

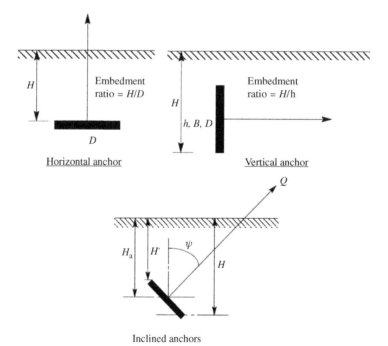

Figure 2.2 Soil anchor plates based on applications by Merifield and Sloan (2006).

cohesionless soils. This type of soil anchor plate can be in various shapes (square, strip, rectangular, or circular plates) in most geotechnical projects. A horizontal anchor plate includes a width (D) and a length (L), where the length must be greater or equal to than the width $(L \geq D)$. The ratio of length to width of the soil anchor plate is called the embedment ratio (L/D). Soil anchor plates subjected to vertical loads are called horizontal anchor plates. These plates can show two types of failure surfaces based on their embedment ratio. The horizontal anchor plates are separated into two types, namely shallow and deep anchor plates. Shallow anchor plate failure extends from the failure surface to ground surface, while the deep anchor plates exhibit the shear force in and around the horizontal anchor plate. If the embedment ratio (L/D) is small, then the failure at the surface of a shallow anchor plate extends to the ground surface, while a large embedment ratio indicates a deep anchor plate. Horizontal anchor plates include a net ultimate uplift capacity (Q_{nu}) and an effective self-weight of the horizontal anchor plate (W). The sum of the net ultimate uplift capacity and effective self-weight of the horizontal anchor plate is equal to the ultimate uplift capacity of soil anchor plates subjected to uplift load (Q_u). The failure surface makes an angle (α), which is the angle between the failure surface and the ground surface as illustrated in Fig. 2.3.

This angle varies in different soils. For example, in soft cohesive soils and loose cohesionless soils, the α may be close to $90°$, while stiff cohesive soils and dense cohesionless soils may be equal to $45°$-$Ø/2$.

$$Q_u = Q_{nu} + W$$

$Ø$ = Soil friction angle
Q_u = ultimate uplift capacity of soil anchor plate subjected to uplift load

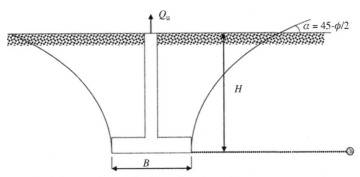

Figure 2.3 The failure surface in horizontal anchor plate.

Q_{nu} = net ultimate uplift capacity = the sum of weight of the soil placed in the failure surface

W = self-weight of the horizontal anchor plate

Later chapters in this book discuss the performance of horizontal anchor plates in cohesive and cohesionless soils.

2.2.1.2 Vertical Anchor Plate

Vertical anchor plates can be used in geotechnical projects to resist horizontal loads, such as at the base of retaining walls, sea walls, sheet pile walls and related projects. The prediction of the ultimate capacities of vertical anchor plates and their displacement are very important in these projects, because the performance of vertical anchor plates, which use tie rods and plates in their construction, subjected to lateral earth pressure is quite different with rigid earth structures (as suggested by Rankine and Coulombs (Rowe, 1952)). A vertical anchor plate includes the height (h), width (D) and embedment depth (H) as illustrated in Fig. 2.3.

Vertical anchor plates (Fig. 2.4) can also be classified into shallow and deep anchor plates. Shallow anchor plates extend their failure surface to ground surface, while the deep anchor plates exert shear force in and around the vertical anchor plate. If the embedment ratio (H/h) is small, then the failure surface of a shallow anchor plate extends to the ground surface, while a large embedment ratio indicates a deep anchor plate. It is important that the passive force is in front of the vertical anchor plate, although the ultimate force capacity (Q_u) is equal to the net ultimate force capacity (Q_{nu}) in vertical anchor plates. Later chapters in this book discuss the performance of vertical anchor plates in cohesive and cohesionless soils.

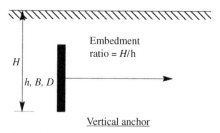

Figure 2.4 Vertical anchor plate.

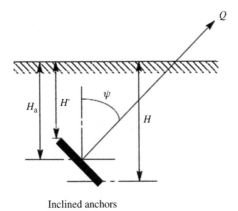

Inclined anchors

Figure 2.5 Inclined anchor plate.

2.2.1.3 Inclined Anchor Plate

Most geotechnical projects need to incline loads. In many projects, such as foundations, transmission towers, pipelines and related projects, inclined anchor plates are sometimes located at an inclination subjected to inclined loads. Most projects subjected to uplift and pullout require inclined anchor plates, although a few projects use inclined anchor plates subjected to axial pull. If axial pull is considered in a project, then the ultimate force capacity can be determined as:

$Q_u = Q_{nu} + W \cos \psi$

ψ = Angle of inclination of anchor plate to the vertical axis, as illustrated in Fig. 2.5.

Q_u = ultimate force capacity of soil anchor plate subjected to pull

Q_{nu} = net ultimate force capacity = the sum of weight of the soil placed in the failure surface

W = self-weight of the inclined anchor plate

2.2.2 Types of Anchor Plates Based on Shape Factor

The shape of an anchor plate is an important factor in the optimum selection of soil anchor plates, as the cost and materials in various geotechnical projects can then also be optimized. All types of soil anchor plate shapes (square, strip, rectangular, and circular) can be used in all geotechnical projects as illustrated in Fig. 2.6, but the size of the materials and the shape are very important considerations in the design and construction. This will be discussed further in later chapters in this book.

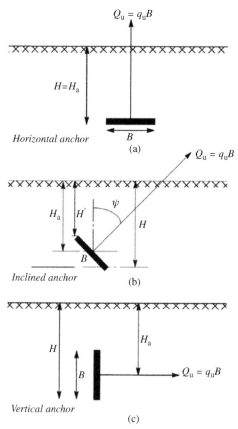

Figure 2.6 Types of anchor plates. (a) Horizontal anchor (b) Inclined anchor (c) Vertical anchor.

2.2.2.1 Square Anchor Plate

The square anchor plate is a type of soil anchor plate that has a height (h) and width (D) that are equal ($h = D$). This soil anchor plate can be fabricated from steel, precast concrete, timber, and fiber reinforced polymer. The square anchor plate includes the square plate and a tie rod. It is used in horizontal, vertical, and inclined systems, as illustrated in Fig. 2.7. These soil anchor plates can be used in group anchor plate systems in various earth projects.

2.2.2.2 Rectangular Anchor Plate

In most projects, the height (h) to width (D) (D/h ratio) is less than 5 in rectangular anchor plates. Rectangular anchor plates are made of steel, precast concrete, timber, and fiber reinforced polymer and include tie rod(s) for installation in various projects as illustrated in Fig. 2.8.

B

Square anchor (*L=B*)

Figure 2.7 Square anchor plates.

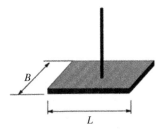

B

L

Rectangular anchor (*B>>L*)

Figure 2.8 Rectangular anchor plates.

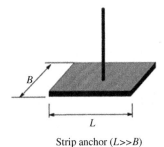

B

L

Strip anchor (*L>>B*)

Figure 2.9 Strip anchor plates.

2.2.2.3 *Strip Anchor Plate*

Strip anchor plates have a height (h) to width (D) (D/h ratio) greater than 6. The material composition of strip anchor plates includes steel, timber, and precast concrete. These anchor plates need tie rod(s) in installation of various earth projects when subjected to inclined, vertical and horizontal loads, as illustrated in Fig. 2.9.

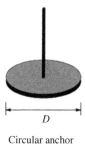

Circular anchor

Figure 2.10 Circular anchor plates.

2.2.2.4 Circular Anchor Plate

A good type of group anchor plate is the circular plate. These anchor plates are made of various materials such as precast concrete, steel, timber, and fiber reinforced polymer. They can be used in various geotechnical projects to resist uplift, horizontal, and inclined loads under axial load or pullout. Fig. 2.10 illustrates circular anchor plates.

2.3 EARLY THEORIES

Most research on this topic has been during the past 50 years or so, but some theories are as early as 1957 (Hueckel, 1957), followed closely by Mors (1959), Downs and Chieurzzi (1966) and other researchers. Based on the application types of soil anchor plates, the early theories were separated according to horizontal anchor plates (Mors's theory, Down and Chieurzzi's theory) and vertical anchor plates (Hueckel's theory).

2.3.1 Cone Method (Mors's theory (1959) and Downs and Chieurzzi's theory (1966))

Mors (1959) showed the relationship between the failure surface under ultimate load and the failure angle using circular anchor plates in shallow conditions. He described a failure angle of $\Theta = 90° + \emptyset/2$ at ultimate uplift capacity and the failure surface as a cone, as illustrated in Fig. 2.11. The net ultimate uplift capacity can be considered to be equal to the weight of soil in the failure surface.

$$Q_u = \gamma V$$

where $V =$ the soil volume in the cone

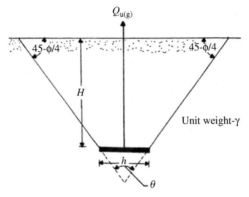

Figure 2.11 Cone method (Mors's theory, 1959).

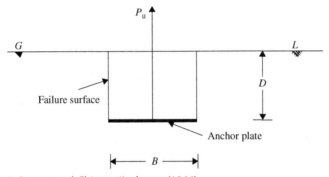

Figure 2.12 Downs and Chieurzzi's theory (1966).

$$V = \frac{\pi}{3}H\left\{ h^2 + \left[h + 2H \cot\left(45 - \frac{\varnothing}{4} \right) \right]^2 + h\left[h + 2H \cot\left(45 - \frac{\varnothing}{4} \right) \right] \right\}$$

$$= \frac{\pi H}{3}\left[3h^2 + 4H^2 \cot^2\left(45 - \frac{\varnothing}{4} \right) + 6H\, h \cot\left(45 - \frac{\varnothing}{4} \right) \right]$$

γ = unit weight of soil

Downs and Chieurzzi (1966) showed a failure angle (Θ) equal to 60° using a circular anchor plate in shallow conditions, as illustrated in Fig. 2.12, demonstrating that the net ultimate uplift capacity is:

$$Q_u = \gamma V = \frac{\pi\gamma H^3}{3}\left[h^2 + [h + 2H \cot 60°]^2 + h(h + 2H \cot 60°) \right]\}$$

$$= \frac{\pi\gamma H^3}{3}\left[3h^2 + 1.33H^2 + 3.64H \right]$$

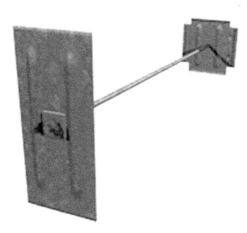

Figure 2.13 Soil anchor plate.

2.3.2 Friction Method

In this method, the net ultimate uplift capacity (Q_u) is determined using a circular anchor plate in shallow conditions, which is considered cylindrical, as illustrated in Fig. 2.13.

$$Q_u(\text{Cohesionless soils}) = \frac{\pi h^2}{4} H\gamma + \int_0^H (\sigma_0' \tan \varnothing) dZ$$

where σ_0' = effective overburden pressure

$$Q_u(\text{Cohesive soils}) = \frac{\pi H h^2 \gamma}{4} + \pi H h C_u$$

where C_u = undrained cohesion

2.4 CONSTRUCTION OF ANCHOR PLATES

Soil anchor plates are made from steel plates, timber sheets, precast concrete slabs, and fiber reinforced polymers. These types of soil anchors can be fabricated into various shapes, such as square, strip, rectangular and circular shapes. The soil anchor plates are strategically placed, based on the design, and are then backfilled to secure them. The first step in the installation of soil anchor plates is the onsite excavation and strategic placement of the anchor plates with their tie rods located in the excavated zone. The components of a soil anchor plate include the anchor plate, tie rod, and the plug (installation of earth structure to tie rod), as illustrated in Fig. 2.13.

The construction steps illustrated in Figs. 2.14−2.16 show the various steps and samples of installation of soil anchor plates.

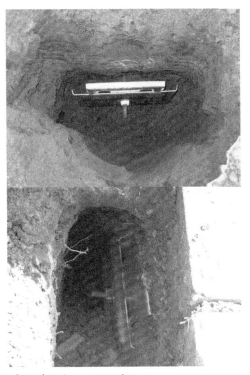

Figure 2.14 Soil anchor plate in construction.

If a system comprising group anchor plates is required, then the center-to-center spacing design of the plates is a very important component of the construction of various geotechnical projects as illustrated in Fig. 2.17. The group anchor plate system capacity may possibly be smaller than the ultimate capacity of a single anchor plate in the system and this can be due to interferences of soil anchor plates within the system.

The new types of soil anchor plates used in reinforced soils can be ideally situated and then geosynthetics can be placed on them afterwards prior to backfilling the soil in the excavated zone to further secure them. The role of geosynthetics is very important in soil reinforcement and in the improved performance of soil anchor plates in practice. Geotextiles, geogrids, and other types of geosynthetics can be used as soil reinforcement materials in reinforced anchor plates, but geomembranes showed a weak effect on the performance of soil anchor plates in various geotechnical projects (Niroumand et al., 2013). The backfill steps include compaction of soil layers on soil anchor plates and geosynthetics in practice as illustrated in Fig. 2.18. Consideration of the number of impacts is also very important when determining the compaction of backfill in various soil layers.

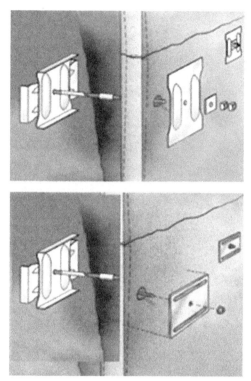

Figure 2.15 Construction steps of soil anchor plate.

Figure 2.16 Soil anchor plates used in foundations.

Figure 2.17 Group anchor plates.

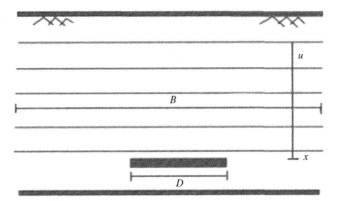

Figure 2.18 Soil anchor plates and geosynthetics.

2.5 CONSTRUCTION PROBLEMS

Excavation, placement of the geosynthetic layers and soil anchor plates, and backfill are the main considerations in using soil anchor plates. Excavation changes the properties and soil conditions in various practical projects, which is a drawback of soil anchor plates. Engineers and researchers have restricted options in this regard, nonetheless backfill with compaction is a very good solution for this problem. The placement of the soil anchor plates and geosynthetic layers is another crucial factor in geotechnical projects and it is very important that geosynthetics improves the performance of soil anchor plates. Consideration of the backfill conditions is extremely important in installation of soil anchor plates in practical projects. It is clear that soil anchor plates serve limited purpose in geotechnical structures such as tunnels, some foundations and related projects that do not allow excavation during construction; this is another limitation of soil anchor plates in most earth structures. The cost of soil anchor plates increases with increasing excavations steps in practice. Soil anchor plates have another construction consideration when implementing them in multiple soil layers For example, in multiple layers of cohesionless soils, the loose sand and dense cohesionless soils do not interact well between two layers. Other projects may have issues with the combination of cohesive and cohesionless soils.

2.6 CONCLUSION

This chapter discussed the definitions of soil anchor plates. Soil anchor plates have their own innate limitations, such as the need to excavate and backfill in various geotechnical projects, however their importance in most geotechnical projects such as retaining walls, sea walls and related projects is crucial. Employing soil reinforcement materials can significantly increase the performance of soil anchor plates in various geotechnical projects.

REFERENCES

Das, B.M., 1990. Earth Anchors. Elsevier, Amsterdam.
Downs, D.I., Chieurzzi, R., 1966. Transmission tower foundations. J. Power Div. ASCE 88 (2), 91−114.
Hueckel, S., 1957. Model Tests on Anchoring Capacity of Vertical and Inclined Plates. Proceedings of Fourth International Conference on Soil Mechanics and Foundation Engineering, London, Vol. 2, pp. 203−206.
Merifield, R., Sloan, S.W., 2006. The ultimate pullout capacity of anchors in frictional soils. Can. Geotech. J. 43 (8), 852−868.
Mors, H., 1959. The behaviour of mast foundations subjected to tensile forces. Bautechnik 36 (10), 367−378.
Niroumand, et al., 2013. The influence of soil reinforcement on the uplift response of symmetrical anchor plate embedded in sand. Meas., Elsevier 46 (8), 2608−2629, October 2013.
Rowe, P.W., 1952. Anchored sheet pile walls, proceeding. Institution of Civil Eng. Vol. 1 (No. 1), 27−70, London, January.

CHAPTER 3

Horizontal Anchor Plates in Cohesionless Soil

3.1 INTRODUCTION

Anchor plates are used in various vertical loads such as uplift loads. Anchor plates come in four shapes: square, circular, rectangular and strip forms, as illustrated in Fig. 3.1. Anchor plates are categorized based on their applications: horizontal, vertical, and inclined anchor plates, and these are shown in Fig. 3.2. Horizontal anchor plates are a combination of plate shapes, rod or tendons and various joints used as an earth anchorage in the ground, as illustrated in Fig. 3.3.

Horizontal anchor plates are a good solution for geotechnical projects that need resistance again vertical loads, eg, a transmission tower or pipeline. These anchors can be classified as either shallow or deep anchor plates, determined by the embedment ratio. The embedment ratio is the ratio of L/D (where L is the embedment depth and D is the plate width/diameter). If the embedment ratio is small, then horizontal anchor plates will be placed in shallow troughs although they may be separated into single and group anchor plates.

Group horizontal anchor plates (Fig. 3.4) are a better solution for big uplift loads.

Horizontal anchor plates can be used in various soils but in this chapter we will discuss use in cohesionless soils only. Cohesionless soils are defined as any free-running type of soil, such as sand or gravel, whose strength depends on friction between particles (measured by the friction angle, Ø). In this chapter we will look at various horizontal anchor plates in cohesionless soils.

3.2 BASIC PARAMETERS

Researchers vary in what they define as the most important aspect of the use of horizontal anchor plates. Looking back at chapter "Anchor Plates,"

Design and Construction of Soil Anchor Plates.
DOI: http://dx.doi.org/10.1016/B978-0-12-420115-6.00003-5

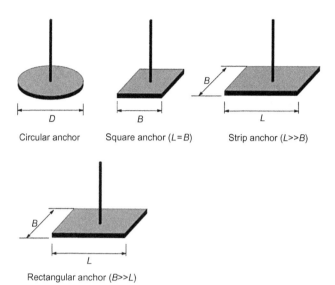

Circular anchor   Square anchor (L=B)   Strip anchor (L>>B)

Rectangular anchor (B>>L)

Figure 3.1 Soil anchor plates based on various shapes.

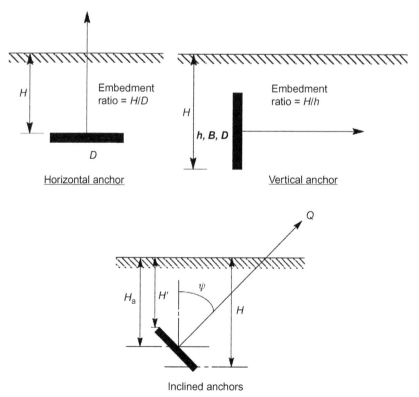

Horizontal anchor   Vertical anchor

Inclined anchors

Figure 3.2 Soil anchor plates based on applications.

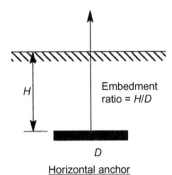

Figure 3.3 Horizontal anchor plate in geotechnical engineering.

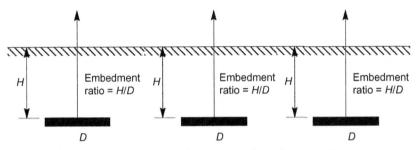

Figure 3.4 Group horizontal anchor plate in geotechnical engineering.

Mors (1959) nominated the importance of embedment depth and plate width in horizontal anchor plates because they could be affected by failure angle. The role of soil volume is so important because it can increase the uplift capacity. The unit weight of soil, related of course to the soil type, is another factor in the pullout capacity of horizontal anchor plates. The basic parameters of horizontal anchor plates in cohesionless soil can defined by

$$Q_u = \gamma V$$

where

Q_u, the net ultimate uplift capacity

V, the soil volume

γ, unit weight of soil.

Most published research has been focused on evaluating horizontal anchor plates in nonreinforced sand, but Niroumand et al. (2013)

evaluated the importance of soil reinforcement in the uplift response of horizontal anchor plates, using points as discussed in this chapter.

3.3 FAILURE MODE

The failure mechanism of horizontal anchor plates in sand differ, based on plate shapes, sand type, soil reinforcement condition and shallow or deep conditions. Cohesionless soil (which can be categorized into loose, medium, and dense conditions) can have an effect on the failure mechanism of horizontal anchor plates. The failure mechanism of horizontal anchor plates are as illustrated in Figs. 3.5 and 3.6.

3.3.1 Niroumand's Method

Niroumand et al. (2013) investigated the failure shape of horizontal anchor plates (circular, rectangular, square, and strip plates) in reinforced

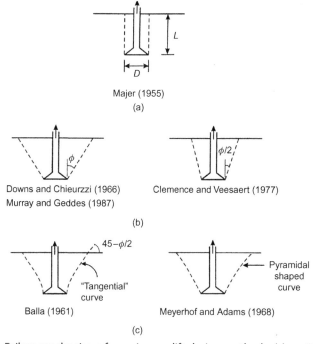

Figure 3.5 Failure mechanisms for various uplift design methods: (a) vertical slip surface model, (b) inverted truncated cone model, and (c) curved slip surface model.

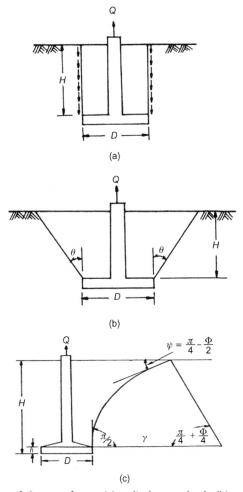

Figure 3.6 Assume failure surfaces: (a) cylinder method, (b) cone method, and (c) Balla's method.

sand by geogrid and grid-fixed reinforced (GFR) as an innovative soil reinforcement system that tied the geogrid to soil. Using the soil reinforcement changed the failure mechanism of horizontal anchor plates in the reinforced zone of soil because it makes two failure zones that need to be considered: the first zone is the relation between the plate anchor and soil reinforcement materials; the second zone between soil reinforcement layer and ground surface.

3.4 UPLIFT CAPACITY FOR HORIZONTAL ANCHOR PLATES

In this section, we will briefly look at the research that has been carried out on various theoretical methods and techniques affecting the pullout capacity of horizontal anchor plates.

3.4.1 Mors's Theory (1959)

Mors (1959) evaluated the failure surface under ultimate load and the failure angle using circular anchor plates in shallow conditions. He described a failure angle of $\Theta = 90° + \varnothing/2$ at ultimate uplift capacity and the failure surface as a cone, as illustrated in Fig. 3.7. The net ultimate uplift capacity can be considered to be equal to the weight of soil in the failure surface.

$$Q_u = \gamma V$$

where V is the soil volume in the cone

$$V = \frac{\pi}{3} H \left\{ h^2 + \left[h + 2H \cot\left(45 - \frac{\varnothing}{4} \right) \right]^2 + h \left[h + 2H \cot\left(45 - \frac{\varnothing}{4} \right) \right] \right\}$$

$$= \frac{\pi H}{3} \left[3h^2 + 4H^2 \cot^2\left(45 - \frac{\varnothing}{4} \right) + 6H\, h \cot\left(45 - \frac{\varnothing}{4} \right) \right]$$

where γ, unit weight of soil.

Methods for estimating the ultimate pullout capacity of plate anchors have been developed. One of the earliest publications concerning ultimate

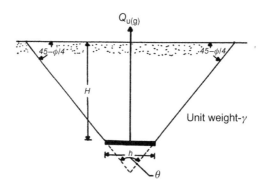

Figure 3.7 Cone method (Mors's theory, 1959).

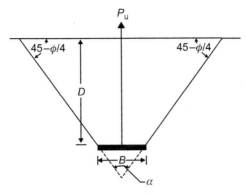

Figure 3.8 Failure surface assumed by Mors (1959).

pullout capacity of anchor plates was by Mors (1959) who proposed a failure surface in the soil at ultimate load which may be approximated as a truncated cone, having an apex angle α equal to $(90° + \varnothing/2)$ as shown in Fig. 3.8. The net ultimate pullout capacity was assumed to be equal to the weight of the soil mass bounded by the sides of the cone and the shearing resistance over the failure area surface was ignored.

$$P_u = \gamma V$$

where

V, volume of the soil in the truncated

γ, unit weight of soil.

This information can be used for the design and evaluation of anchor systems used to prevent the sliding and/or overturning of laterally loaded structures founded in soils. The typically system of forces acting on a simple anchor is shown in Fig. 3.9. The pullout force is given by typical equation:

$$P_u = P_s + W + P_t$$

where

P_u, ultimate pullout force

w, effective weight of soil located in the failure zone

P_s, shearing resistance in failure zone

P_t, force below the area.

In this case of sands P_t is equal to zero.

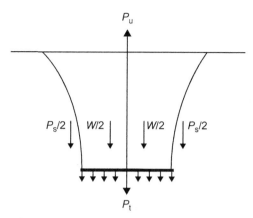

Figure 3.9 Anchor plate under pullout test.

3.4.2 Friction Method

In this method the net ultimate uplift capacity (Q_u) is determined using a circular anchor plate in shallow conditions, where the failure shape is considered to be cylindrical.

$$Q_u(\text{Cohesionless soils}) = \frac{\pi h2}{4} H\gamma + \int_0^H (\sigma'_0 tan\ \emptyset)dZ$$

where σ'_0, effective overburden pressure.

3.4.3 Balla's Method

Subsequent variations upon these early theories have been proposed, such as the proposal of Balla (1961), who determined the shape of the slip surfaces for the horizontal anchor plates in sand as illustrated in Fig. 3.10. He proposed equations for estimating the force of the anchors based on the observed shapes of the slip surfaces. He proposed a method to predict the ultimate pullout capacity of anchor plate. Balla developed a shearing resistance equation during failure surface:

$$P_u = L^3\gamma\left[F_1\left(\emptyset, \frac{L}{D}\right) + F_3\left(\emptyset, \frac{L}{D}\right)\right]$$

The sum of the functions F_1, F_2 can be obtained by Fig. 3.11. The breakout factor N_q is defined as

$$N_q = \frac{P_u}{\gamma AL}$$

where A, area of plate.

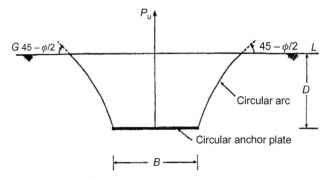

Figure 3.10 Failure surface on circular anchor plate assumed by Balla (1961).

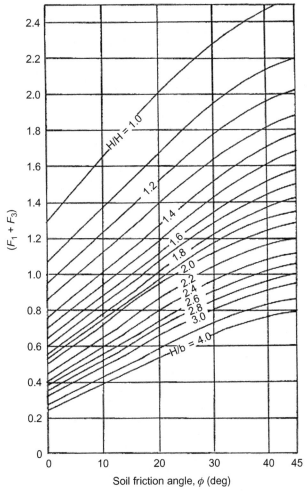

Figure 3.11 Variation of $F_1 + F_3$ based on Balla's result (1961).

The breakout factor increases with L/D up to N_q^* at $L/D_{(cr)}$. Based on Balla's method, the shallow anchor plates are defined at $L/D \leq L/D_{(cr)}$ and deep anchor plates at $L/D \geq L/D_{(cr)}$.

3.4.4 Baker and Kondner's Method

Baker and Kondner (1966) approved Balla's method findings regarding the behavioral difference between the deep and shallow anchors in dense sand. They suggested the below equations:

$$\text{Shallow circular anchor plate,} \quad Q_u = C_1 \, LD^3\gamma + C_2 \, D^3\gamma$$

$$\text{Deep circular anchor plate,} \quad Q_u = 170 \, D^3\gamma + C_3 \, D^2 t\gamma + C_4 LD$$

where t, the thickness of anchor plate and C_1, C_2, C_3, C_4 are constants based on soil friction angle and relative density of compaction.

3.4.5 Mariupol'skii's Method

Mariupol'skii (1965) investigated the mathematical equations related to the uplift response of circular anchor plates. He evaluated the uplift performance in shallow and deep anchor plates. He assumed the initial force based on the effective weight of the anchor plate, the effective weight of the soil column of diameter (D) and height (L), and the friction and cohesion along the column. He suggested the uplift capacity of anchor plate for shallow anchor plates can be calculated as

$$Q_u = \frac{\pi}{4}(D^2 - d^2)\frac{\gamma L(1 - (d/D)^2 + 2K_0(L/D)\tan \varphi) + 4c(L/D)}{1 - (L/D)^2 - 2n(L/D)}$$

where

K_0, lateral earth pressure coefficient
C, cohesion (although it is 0 for sand)
n, $0.025\varnothing$
d, diameter of anchor rod.

He also proposed an equation for deep circular anchor plates:

$$Q_u = \frac{\pi q_0}{2}\left[\frac{D^2 - d^2}{\tan \varphi}\right] + f(\pi d)[L - (D - d)]$$

where q_0, radial pressure under which the cavity is expanded and f, unit skin friction along the stem of the anchor plate.

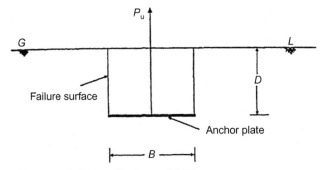

Figure 3.12 Downs and Chieurzzi's theory (1966).

3.4.6 Downs and Chieurzzi's Theory (1966)

Downs and Chieurzzi (1966) showed a failure angle (Θ) equal to $60°$ using a circular anchor plate in shallow conditions, as illustrated in Fig. 3.12, demonstrating that the net ultimate uplift capacity is

$$Q_u = \gamma V = \frac{\pi \gamma H^3}{3} \left\{ \left[h^2 + [h + 2H \cot 60°]^2 + h(h + 2H \cot 60°) \right] \right\}$$

$$= \frac{\pi \gamma H^3}{3} \left[3h^2 + 1.33H^2 + 3.64H \right]$$

3.4.7 Meyerhof and Adam's Method

An approximate semi-empirical theory for the pullout loading force of horizontal strip, circular, and rectangular anchors was proposed by Meyerhof and Adams (1968). For a strip anchor, an expression for the ultimate pullout capacity was selected by considering the equilibrium of the block of soil directly above the anchor (ie, contained within the zone made when vertical planes are extended from the anchor edges). The capacity was assumed to act along the vertical planes extending from the anchor shape, while the total passive earth pressure was assumed to act at some angle to these vertical planes. This angle was selected based on laboratory test results while the passive earth pressures were evaluated from the results of Caquot and Kerisel (1948). For shallow plate anchors where the failure surface develops at the soil surface, the ultimate pullout capacity was determined by considering equilibrium of the material between the anchor and soil surface. For a deep anchor, the equilibrium of a block of soil extending a vertical distance (H) above the anchor was presented, where H was less than the actual embedment depth of the plate anchor. The magnitude of H was determined from the observed extent of the failure surface from laboratory works. The analysis of strip footings was developed by Meyerhof and Adams to include circular plate anchors by using a semi-empirical shape factor to modify the passive earth pressure obtained for the plane strain case. The failure surface was assumed to be a vertical cylindrical surface through the anchor edge and extending to the soil surface.

An approximate analysis for the capacity of rectangular plate anchors was selected for downward loads (Meyerhof, 1951), by assuming the ground pressure along the circular perimeter of the two end portions of the failure surface was governed by the same shape factor assumed for circular anchors. It was, however, based on two key features: the edge of the failure surface and the distribution of stress along the failure surface. Even so, the theory presented by Meyerhof and Adams (1968) has been found to give reasonable estimates for a wide range of plate anchor problems. It is one of only two methods available for appraising the force of rectangular plate anchors. Meyerhof and Adams (1968) expressed the ultimate pullout capacity in rectangular anchor plates as

$$P_u = W + \gamma H^2 (2S_f L + B - L) K_u \tan \varnothing$$

$$S_f = 1 + m \frac{L}{D}$$

$$N_q = 1 + \frac{L}{D} K_u \tan \varnothing$$

where K_u, the nominal uplift coefficient based on Fig. 3.13.

S_f, the shape factor

W, the weight of anchor plate

m, the coefficient that is a function of the soil friction angle based on Fig. 3.14.

They proposed the following equations for strip anchor plate calculations:

$$S_f = 1 + m \frac{L}{D}$$

$$N_q = 1 + K_u \left(\frac{L}{D}\right) \tan \varnothing$$

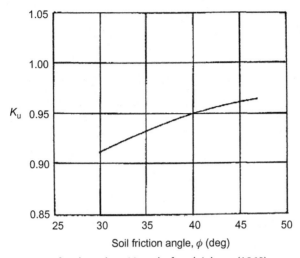

Figure 3.13 Variation of K_u based on Meyerhof and Adams (1968).

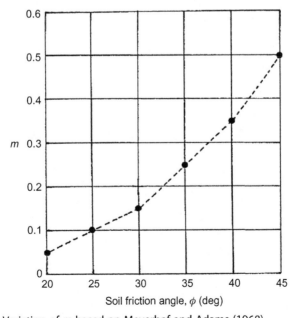

Figure 3.14 Variation of m based on Meyerhof and Adams (1968).

Meyerhof and Adams (1968) suggested the below formula for circular anchor plates:

$$Q_u = W + \frac{\pi}{2} S_f \gamma D L^2 K_u \tan \varnothing$$

$$S_f = 1 + m \frac{L}{D}$$

$$N_q = 1 + 2 \left[1 + m \left(\frac{L}{D} \right) \right] \left(\frac{L}{D} \right) K_u \tan \varnothing$$

3.4.8 Vesic's Method

Vesic (1971) studied the problem of an explosive point charge expanding a spherical area close to a surface of semi-infinite, homogeneous and isotropic soils as illustrated in Fig. 3.15. He assumed the vertical component of the force inside the cavity (P_v), effective self-weight of the soil ($W = W_1 + W_2$), and vertical component of the resultant of internal force (F_v). Fig. 3.16 illustrates the breakout factor of strip plates under pullout load in sand based on Vesic's method.

$$P_u = \gamma H N_q$$

$$N_q = \left[1 + A_1 \left(\frac{H}{h_1/2} \right) + A_2 \left(\frac{H}{h_1/2} \right)^2 \right]$$

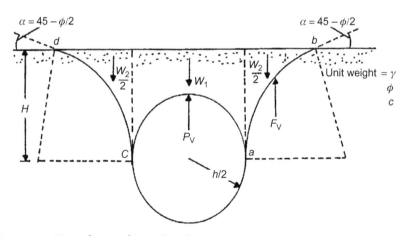

Figure 3.15 View of tests of Vesic (1971).

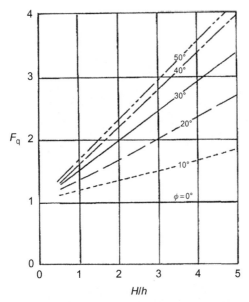

Figure 3.16 Breakout factor in strip anchor plate of Vesic (1971).

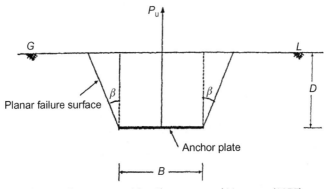

Figure 3.17 Failure surface assumed by Clemence and Veesaert (1977).

3.4.9 Clemence and Veesaert Method

Clemence and Veesaert (1977) showed a formulation for shallow circular anchors in sand assuming a linear failure making an angle of $\beta = \varnothing/2$ with the vertical through the shape of the anchor plate as shown in Fig. 3.17. The contribution of shearing resistance along the length of failure surface was approximately taken into consideration by selecting a suitable value of ground pressure coefficient from laboratory model works. The net ultimate pullout capacity can be given as

$$P_{u} = \gamma V + \pi \gamma K_{o} \tan \varnothing \cos^2\left(\frac{\varnothing}{2}\right)\left(\frac{BD^2}{2} + \frac{D^3 \tan(\varnothing/2)}{3}\right)$$

where

V, the volume of the truncated cone above the anchor

K_{o}, the coefficient of lateral earth pressure, they suggested that the magnitude of K_{o} may vary between 0.6 and 1.5 with an average value of about 1.

3.4.10 Rowe and Davis's Method

Rowe and Davis (1982) presented a research of the behavior of anchor plate in sand. Tagaya et al. (1983, 1988) conducted two-dimensional plane strain and axisymmetric finite element analyses using the constitutive law of Lade and Duncan (1975). Scale effects for circular plate anchors in dense sand were investigated by Sakai and Tanaka (1998) using a constitutive model for a nonassociated strain hardening-softening elastoplastic material. The finite element method (FEM) had also been used by Vermeer and Sutjiadi (1985), Tagaya et al. (1983, 1988), and Sakai and Tanaka (1998). Unfortunately, only limited results were presented in these groups of research. The effect of shear band thickness was also introduced. They investigated the relationship between various variables of anchor plates as Fig. 3.18.

3.4.11 Koutsabeloulis and Griffiths's Method

Koutsabeloulis and Griffiths (1989) investigated the trapdoor problem using the initial stress FEM. Both plane strain and axisymmetric research were conducted. The researchers concluded that an associated flow rule has little effect on the collapse load for strip plate anchors but a significant effect (30%) for circular anchors. Large displacements were observed for circular plate anchors prior to collapse. In the limit equilibrium method (LEM), an arbitrary failure surface is adopted along with a distribution of stress along the selected surface. Equilibrium conditions are then considered for the failing soil mass and an estimate of the collapse load is assumed. In the study of horizontal anchor force, the failure mechanism is generally assumed to be log spiral in edge (Saeedy, 1987; Sarac, 1989; Murray and Geddes, 1987; Ghaly and Hanna, 1994b) and

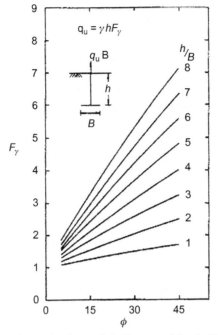

Figure 3.18 Variation of capacity factor F_γ in Rowe and Davis (1982).

the distribution of stress is obtained by using either Kotter's equation (Balla, 1961), or by using an assumption regarding the orientation of the resultant force acting on the failure plane. The equation of Murray and Geddes (1982) is

$$N_q = 1 + \frac{L}{D}\tan\varnothing\left(1 + \frac{D}{B} + \frac{\pi L}{3 B}\right)$$

Upper and lower bound limit analysis techniques have been studied by Murray and Geddes (1987, 1989), Basudhar and Singh (1994), and Smith (1998) to estimate the capacity of horizontal and vertical strip plate anchors. Basudhar and Singh (1994) selected estimates using a generalized lower bound procedure based on finite elements and non-linear programming similar to that of Sloan (1988). The solutions of Murray and Geddes (1987, 1989) were selected by manually constructing cinematically admissible failure mechanisms (upper bound), while Smith (1998) showed a novel rigorous limiting stress field (lower bound) solution for the trapdoor problem.

3.4.12 Remeshbabu's Method

Ghaly (1977) had recommended a general expression for the pullout capacity of the vertical anchor plates based on the analysis of the experimental test results from the published literature. Along similar lines and incorporating appropriate correction, Ramesh babu (1998) proposed a general expression for the horizontal anchor plates in the sand by analyzing the results of published experimental data and his own pullout tests data. For a horizontal strip horizontal anchor plate:

$$\frac{P_u}{\gamma A D \tan \emptyset} = 3.24 \left(\frac{D^2}{A}\right)^{0.34}$$

For square and circular horizontal anchor plates:

$$\frac{P_u}{\gamma A D \tan \emptyset} = 3.74 \left(\frac{D^2}{A}\right)^{0.34}$$

where $P_u/\gamma A D \tan \emptyset$, pullout capacity factor and D^2/A is geometry factor.

3.4.13 Frydman and Shaham's Method

Frydman and Shaham (1989) performed a series of pullout tests on prototype slabs placed at various inclinations and different depths in dense sand. A simple semi-empirical expression is found to reasonably predict the pullout capacity of the continuous, horizontal slab as a foundation for depth−width ratio in their tests. Factors that account for the shape and the inclination are then made, leading to expressions for the estimation of the pullout capacity of any slab anchor. The following expressions have been proposed for the pullout capacity of the horizontal, rectangular slab horizontal anchor plate in dense sand:

$$(N_q)_r = \left[1 + \frac{D}{B} \tan \emptyset\right]\left[1 + \frac{((B/L)-0.15)}{(1-0.15)} \times \left(0.51 + 2.35 \log\left(-\frac{D}{B}\right)\right)\right]$$

For loose sand, $D/B \geq 2$

$$(N_q)_r = \left[1 + \frac{D}{B} \tan \emptyset\right]\left[1 + 0.5\frac{((B/L)-0.15)}{(1-0.15)}\right]$$

where $(N_q)_r$, the pullout capacity.

3.4.14 Related Works on Anchor Plates in Sand

Only a few investigations into the performance of ultimate pullout loading in sand were recorded in model numerical studies. An example of this is Fargic and Marovic (2003) who discussed the pullout capacity of plate anchors in soil under applied vertical force. Computation of the pullout and uplift force was performed by the use of the FEM. For a gravity load, the concept of initial stresses in Gauss points was selected. In the first increment of computation, these stresses were added to the vector of total stress. The soil was modeled by an elastoplastic constituent material model and the associated flow rule was used. The soil mechanics parameters of samples were determined by standard tests conducted on disturbed samples. For a complex constitutive numerical model of material to describe an actual state of soil, a greater number of soil mechanics parameters must be available.

The tensile strength of the soil materials was crucial only in few cases, and the problem of tensile plate anchors is one of them. An iterative procedure was used as the first procedure. The elements with tensile stresses were excluded from the following steps by diminishing the different modulus. More sophisticated constitutive laws are required for an exact analysis, and an adequate FEM code program has to be prepared. Merifield and Sloan (2006) used many numerical solutions for analysis of plate anchors as illustrated in Figs. 3.19−3.21. Until this time very few rigorous numerical analyses had been performed to determine the pullout capacity of plate anchors in sand. Although it is essential to verify theoretical solutions/numerical analysis with experimental studies wherever possible, results selected from their laboratory testing alone were typically problem-specific. This was particularly the case in geotechnical areas, where the researchers were dealing with a highly nonlinear material that

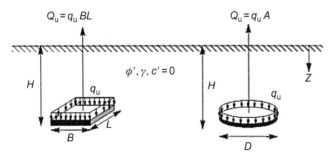

Figure 3.19 Problem definition by Merifield and Sloan (2006).

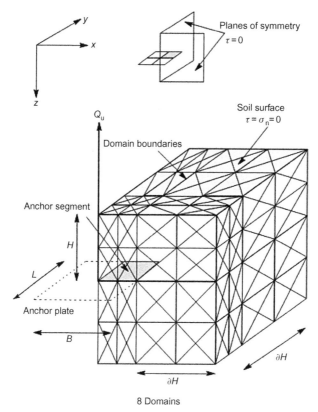

Figure 3.20 Mesh for square anchor plate by Merifield and Sloan (2006).

sometimes displays pronounced scale effects. As the cost of performing laboratory works on each and every field problem combination is prohibitive, it is necessary to be able to model soil pullout loading numerically for the purposes of design. Existing numerical analyses generally assumed a condition of plane strain for the case of a continuous strip plate anchor or axisymmetry for the case of circular plate anchors. The researchers were unaware of any three-dimensional numerical analyses to ascertain the effect of plate anchor shape on the uplift capacity.

Dickin and Laman (2007) investigated the numerical modeling of plate anchors using PLAXIS (a finite element software) as illustrated in Figs. 3.22 and 3.23. Numerical analysis research investigated the uplift response of 1-m-wide strip anchors in sand. The results indicated that the maximum ultimate pullout capacity increased with anchor embedment ratio and sand packing. The research was carried out using a plane strain

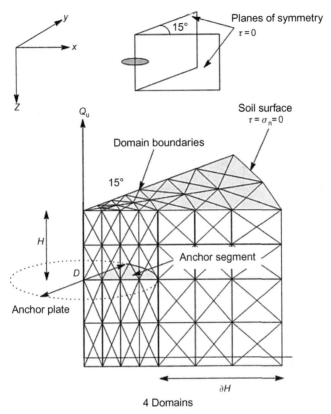

Figure 3.21 Mesh for circular anchor plate by Merifield and Sloan (2006).

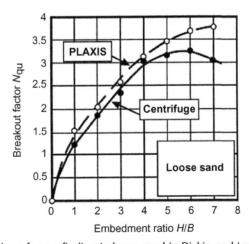

Figure 3.22 Breakout factors finding in loose sand in Dickin and Laman (2007).

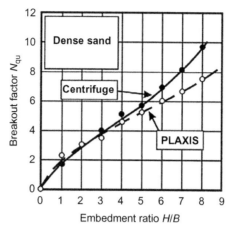

Figure 3.23 Breakout factors finding in dense sand in Dickin and Laman (2007).

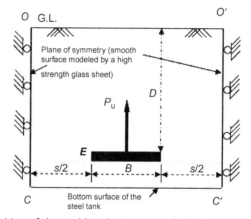

Figure 3.24 Definition of the problem by Kumar and Bhoi (2008).

model for anchors in both loose and dense sand. During the generation of the mesh, 15-node triangular elements were obtained in the determination of stresses.

Kumar and Bhoi (2008) used a group of multiple strip plate anchors placed in sand and subjected to equal magnitudes of vertical upward pullout loads to determine numerical solutions as illustrated in Fig. 3.24. Instead of using a number of anchor plates in numerical modeling, a single plate anchor was used by modeling the boundary conditions along the plane of symmetry on both the sides of the plate anchor. The effect of interference due to a number of multiple strip plate anchors placed in

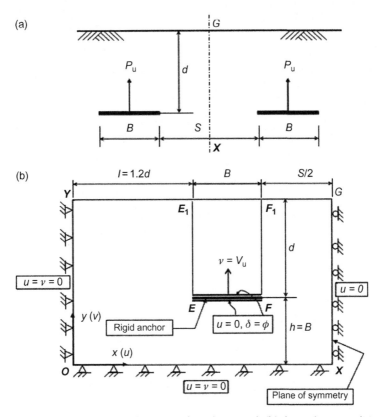

Figure 3.25 (a) Position and loading of anchors and (b) boundary condition by Kuzer and Kumar (2009).

a granular medium at different embedment depths was investigated by conducting a series of small numerical modeling.

Kuzer and Kumar (2009) used a group of two spaced strip plate anchors as illustrated in Fig. 3.25. They investigated the vertical pullout loading of two interfering rigid rough strip anchors embedded horizontally in sand. The analysis was performed by obtained an upper bound theorem of limit analysis combined with finite element and linear programming. The authors used an upper bound finite element limit analysis; the efficiency factor $\xi\gamma$ was computed for a group of two closely spaced strip plate anchors in sand.

Sutherland (1965) showed results for the pullout capacity of the 150 mm horizontal anchors in loose and dense sand, as well as large diameter shafts in medium dense and dense sand. It was concluded that the mode of failure differed according to the sand density and Balla's analytical approach may give reasonable results only in the sand characterized by intermediate density. Kananyan (1966) showed the results for the horizontal circular plate anchors in the loose sand and the medium dense sand. He also performed a series of tests on the inclined anchors and studied the failure on the surface, concluding that most of the soil particles above the anchor moved predominantly in a vertical direction. In these tests, the ultimate uplift load increased in accordance to the inclination angle of the anchors.

Sergeev and Savchenko's (1972) tests were made on few models of AP-1 and AP-2 wooden anchor plates with the dimensions of 74 cm × 74 cm and 60.5 cm × 60.5 cm and thickness of 5 and 10 cm, respectively (shown in Figs. 3.26 and 3.27). The pressure was measured by gages and arranged one next to the other, converting the entire contact surface of the plate into a continuous metering device. Thirty-six gages

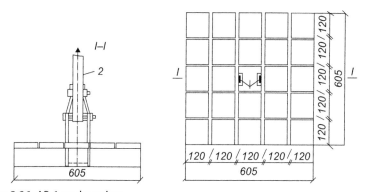

Figure 3.26 AP-1 anchor plate.

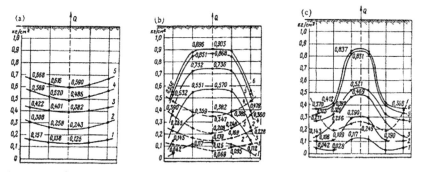

Figure 3.27 Responses pressures: (a) AP-2 in loam, (b) AP-2 in sand, and (c) AP-1 in sand.

were placed on the AP-1 plate, 25 on the AP-2 plate. Readings of resistance gages attached to deformable plastic pressure gages were recorded automatically by an AI-1M strain meter. Gages with two indices of compliance of the rigid plate were used in the experiments. In the more compliant gages, with a cross section of bending plates of 20 mm×3.2 mm, the displacements amounted to 3.2−3.3 mm per kg of loading, and the less compliant gages with a cross-section of the plates of 20 mm×4 mm, these displacements were 1.8−1.9 mm. In most of the experiments, the less compliant gages were placed about the perimeter of the plate model, which corresponded to the character of the theoretical saddle-shaped stress−strain curves obtained in solving the two-dimensional problem. The experiments were conducted in sandy and loamy soil.

Prior to the experiments, in order to eliminate the effects of moisture, the model of the plate was wrapped in two layers of polyethylene sheets and heavy cloth, and it was then set in a predesigned foundation pit with a cross sectional area of 3.5 m×3.5 m and a depth of 2 m. Medium-grain sand was then placed in the pit or loam. Density was controlled by choice of sample cylinder. The unit weight of the sand in unconsolidated state was 1.48 g/cm^3, and in the compacted state 1.65 g/cm^3. The moisture content of the unconsolidated sand was 4.8%, compared to the compacted sand of 4.2%. The standard angle of internal friction of compacted sand in the pressure interval 1−3 kg/cm^2 amounted to 34°; the temerity parameter was 0.02 kg/cm^2, and the strain modulus was $E_0 = 210$ kg/cm^2. The bulk weight of loam with layer-by-layer compaction was 1.90 g/cm^3; the moisture content was in the interval 20−23%. The plastic limit and liquid limit lay in the intervals 21−22% and 35−37%. The standard angle of internal friction was 17°, the cohesion 0.28 kg/cm^2, and the strain modulus $E_0 = 120$ kg/cm^2.

In all experiments with an AP-2 plate, both in unconsolidated sand and in loam, the curves of reaction pressure in the soil have saddle-shaped forms at all stages of loading, including limit load. Repeated loading of the plate introduced no substantial changes in the nature of the distribution of reaction pressures of friable sand or loam. With increased depth of installation of the AP-2 plate in friable sand or loam, the saddle-shaped pressure curves of the soil smoothed out somewhat, which is in qualitative agreement.

Das and Seeley (1975a,b) performed uplift tests for the horizontal rectangular anchors ($L/B \leq 5$) in dry sand with a friction angle of $\emptyset = 31°$ at a density of 14.8 kN/m^3. For each aspect ratio (L/B), it was found that the anchor capacity increases with the embedment ratio before reaching a constant value at the critical embedment depth. A similar investigation was conducted by Rowe (1978) in dry sand with friction angles $\varnothing = 31-33°$ and dry unit weight of $\gamma = 14.9 \text{ kN}/m^3$. Polished steel plates were used for the anchors and the interface roughness was measured as $\delta = 16.7°$. Most tests were performed on the anchors with an aspect ratio L/B of 8.75. Rowe inferred that decreasing the aspect ratio (L/D) leads to the increase of the anchor force (relative to L/B 8.75) of 10%, 25%, 35%, and 120% for L/B ratios of 1−5, respectively. Thus, the effect of the shape is significant for $L/B \le 2$ and is of little importance for $L/B > 5$. This suggests that the anchors with aspect ratios of $L/B > 5$ effectively behave as a continuous strip and can be compared with the methods which obtain the plane strain conditions. In contrast to the observations of Das and Seeley (1975a,b), Rowe (1978) did not observe a critical embedment depth and the anchor capacity was found to be continually increasing with the embedment ratio over the range of $H/B = 1-8$.

Extensive chamber testing programs have been studied by Murray and Geddes (1987, 1989), who performed the pullout load tests on horizontal strip, circular, and rectangular horizontal anchor plates in dense and medium dense sand with $\varnothing = 43.6°$ and $\varnothing = 36°$ respectively. Anchors were typically 50.8 mm in width or diameter and were tested at aspect ratios (L/B) of 1, 2, 5, and 10. Based on the observations, Murray and Geddes reported several conclusions:

1. The pullout load force of rectangular horizontal anchor plates in very dense sand increases with the embedment ratio and with decreasing aspect ratio L/B.
2. There is a difference in the roles for the force of horizontal anchors with rough surfaces compared to those with area smooth surfaces (as much as 15%).
3. Experimental results suggest that the behavior of a horizontal anchor plate with an aspect ratio of $L/B = 10$ is similar to that of the strip type and does not differ much from an anchor with $L/B = 5$.
4. The force of the circular horizontal anchor plates in very dense sand is approximately 1.26 times the capacity of square anchors.

These conclusion confirmed all of Rowe's (1978) findings. It is also of high interest to note that for all the tests performed by Murray and Geddes, no critical embedment depth was seen. More recently, Pearce (2000) performed a series of laboratory uplift tests on the horizontal circular plate anchors, which were pulled vertically in dense sand. These tests were conducted in a large chamber box that was 1 m high and with a diameter of 1 m. Various parameters such as the horizontal anchor plate's diameter, uplift rate and elasticity of the loading system were part of the investigation. The model horizontal anchor plates used for the pullout tests varied in diameter from 50 to 125 mm and were con-structed from 8 mm mild steel. Large diameter anchors were selected (compared with previous research) due to the identified influence of scale effects on the breakout factor for the anchors of diameters less than 50 mm (Andreadis et al., 1981). Dickin (1988) performed 41 tests on 25 mm horizontal anchor plates with aspect ratios of $L/B = 1-8$ at embedment ratios H/B up to 8 in both loose and dense sand. A number of conventional gravity tests were also performed and compared to the centrifuge data. This comparison revealed a significant difference between the prediction for the horizontal anchor plate forces, particu-larly for the square anchors where the conventional test output gave the anchor forces up to twice of that given by the centrifuge. Without explaining the reason, Dickin concluded that the direct extrapolation of the conventional chamber box test into a field scale would provide overoptimistic predictions of the ultimate force for rectangular horizontal anchor plates in the sand.

Tagaya et al. (1988) also performed centrifuge testing on rectangular and circular horizontal anchor plates, although the study was limited when compared to Dickin (1988) mentioned above. Dickin (1988) stud-ied the influence of the anchor geometry, embedment depth and the soil density on the pullout capacity of one-meter prototype horizontal anchor plate, by subjecting 25 mm models to an acceleration of $40\,g$ in a Liverpool centrifuge. It was found that for the strip anchors, pullout resis-tance (expressed as dimensionless breakout factor), increases significantly with the anchor embedment depth and soil density. However, this resistance reduces with the increase in value of the embedment ratio. Failure displacements also increase with the embedment depth but reduce with the soil density and aspect value ratio.

Ramesh babu (1998) investigated the pullout capacity and the load deformation behavior of the horizontal shallow anchor plate. Laboratory experiments were conducted on anchors of different shapes (square, circular, and strip) and embedded in medium dense and dense sands.

In addition, the effect of submergence of the soil above horizontal anchor plates has been investigated. Murray and Geddes' (1996) results

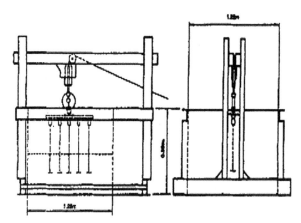

Figure 3.28 Details of pullout test by Murray and Geddes (1996).

are presented of model-scale vertical pulling tests carried out on groups of square anchor row and square configurations. The tests were carried out at a single depth of embedment, with plates in shallow anchor positions in the sand, placed at a constant dry density as illustrated in Figs. 3.28–3.30. It is shown that the load displacement may be reduced to a common curve. The load-carrying capacity of a group of anchor plates increases with the spacing between the individual plates up to a limiting critical value, and the results of pulling tests with different numbers of plates in a group may be demonstrated a simple unifying manner. A possible means of predicting the effect of interaction on the uplift capacity of both is suggested. For laboratory tests on a linear group of five model and full-scale anchors in row configurations anchors, it is shown that the end anchors attain the highest loads but all loads converge to an equal value as the spacing increases to the critical value.

Only a few investigations concerning the performance of the ultimate pullout load in cohesion, have been modeled in the laboratory. An example of this is Fargic and Marovic (2003) who discuss the pullout capacity of the anchors in the soil under the applied uplift force. In the field tests, the pullout forces were gradually increased and the earth surface displacements were measured in two profiles, which were perpendicular to each other, as illustrated in Fig. 3.31. Both the laboratory tests and the field tests were performed for several embedment depths of the sand of the horizontal anchor plate, diameter ratios in the same sand, and under the same conditions.

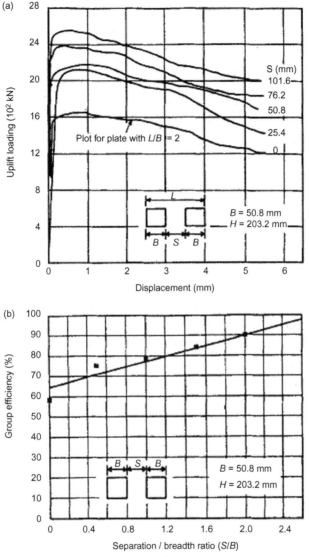

Figure 3.29 Two square anchor plates results from Murray and Geddes: (a) load–displacement curves and (b) group efficiency results.

Murray and Geddes (2006) investigated into the vertical pullout of the horizontal anchor plates in medium dense sand as illustrated in Fig. 3.32. The investigation involved the factors in relation to the load displacement including: the size and shape of plate, depth of embedment, sand density and plate surface roughness. The significant differences in behavior were

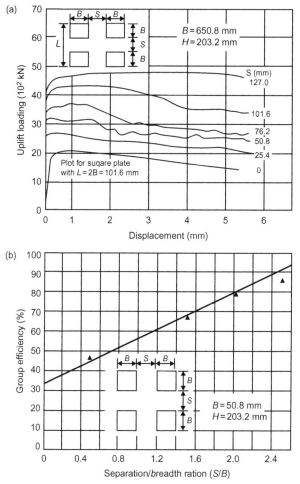

Figure 3.30 Four square anchor plates results from Murray and Geddes: (a) load–displacement curves and (b) group efficiency results.

noted between horizontal anchor plates embedded in very dense sand and those embedded in medium dense sand. The work described within this research is part of a study made of the passive resistance of anchorages in sand. The results of laboratory tests are reported, and comparisons are made with previously published theoretical solutions and equilibrium and limit analysis solutions as developed within this part. The main conclusions of the experimental work were as follows:

1. For the uplift of rectangular plates in very dense sand, the dimensionless load coefficient P/AH and the corresponding displacement at failure,

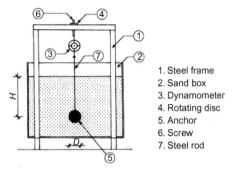

1. Steel frame
2. Sand box
3. Dynamometer
4. Rotating disc
5. Anchor
6. Screw
7. Steel rod

Figure 3.31 Scheme of laboratory test for pullout test by Fargic and Marovic (2003).

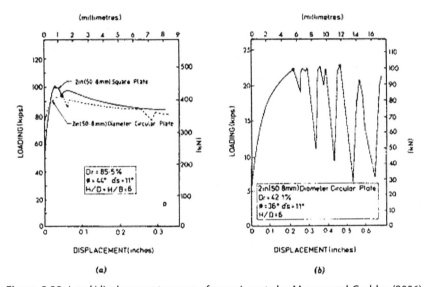

Figure 3.32 Load/displacement curves of experiments by Murray and Geddes (2006): (a) very dense sand and (b) medium dense sand.

increase with an increase of H/B ratio and a decrease of L/B ratio. The dependence of uplift resistance on L/B is described in terms of a shape factor. There is also a marked increase in $P/\gamma\ AH$ and the corresponding displacement in very dense sand for plates with high surface friction angle compared to plates with polished surfaces.

2. Significant differences in behavior were noted between plates embedded in very dense sand and those embedded in medium dense sand. While the dimensionless load coefficient $P/\gamma\ AH$ is greater in very dense sand, the corresponding displacements are considerably

less. Of particular concern is the recorded behavior of circular plates in medium dense sand where large abrupt decreases in uplift resistance were recorded prior to the absolute maximum uplift resistance.

3. For circular plates in very dense sand, there appears to be a consistent relationship for both the dimensionless load coefficients $P/\gamma\, AH$ and the corresponding displacements, for all plates tested, when plotted against H/D. Similar relationships do not appear to exist in medium dense sand. In very dense sand, the $P/\gamma AH$ values for circular plates are, on average, approximately 1.26 times those of square plates for $H/B = H/D$.

Dickin and Laman (2007) investigated the physical modeling of horizontal anchor plates in the centrifuge. The centrifuge incorporates balanced swinging buckets, which were approximately 0.57 m long, 0.46 m wide, and 0.23 m deep. Physical research investigated the pullout response of 1-m-wide strip anchors in sand. Results indicated that maximum resistance increases with the anchor embedment ratio and sand packing. The research showed that the breakout factors for 1-m-wide strip anchors increased with the anchor embedment ratio and the sand packing.

In Dickin (1988), the influence of anchor geometry, embedment, and soil density on the uplift capacity of 1 m prototype anchors is investigated by subjecting 25 mm models to an acceleration of $40\,g$ in the Liverpool centrifuge as illustrated in Fig. 3.33. Uplift resistances, expressed as dimensionless breakout factors, increased significantly with anchor embedment and soil density but reduced with increased aspect ratio. Failure displacements also increased with embedment but reduced with increased soil density and aspect ratio.

The influence of anchor geometry is relatively insensitive to anchor size but increases with both embedment ratio and soil density. In general, the design approaches considered the observed capacity of 1 m strip anchors in dense sand, although predictions using the Meyerhof and Adams and Rowe and Davis theories are acceptable. However, both these approaches appear overoptimistic for anchors in looser sand. In this case, Vesic's theory gives the closest agreement, while the formula of Majer (1955) yields overly conservative designs. Pullout capacities for isolated anchors may be obtained from the strip values in combination with the empirical shape factors reported in this research.

The ultimate uplift resistance of a group of multiple strip anchors placed in sand and subjected to equal magnitudes of vertical upward

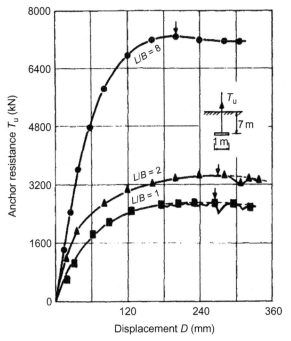

Figure 3.33 Effect of anchor plates in dense sand by Dickin (1988).

pullout loads has been determined by means of model experiments. In Kumar and Kouzer (2008) instead of using a number of anchor plates in the experiments, a single anchor plate was used by simulating the boundary conditions along the planes of symmetry on both the sides of the anchor plate as illustrated in Figs. 3.34 and 3.35. The effect of clear spacing(s) between the anchors, for different combinations of embedment ratio (λ) of anchors and friction angle ($\emptyset$) of soil mass, was examined in detail. The results were presented in terms of an one-dimensional efficiency factor which was defined as the ratio of the failure load for an intervening strip anchor of a given width (B) to that of a single strip anchor plate having the same width. It was clearly noted that the magnitude of efficiency factor reduces quite extensively with a decrease in the spacing between the anchors. The magnitude of efficiency factor for a given s/B was found to vary only marginally with respect to changes in $\emptyset$ and λ. The experimental results presented in this study compare reasonably well with the theoretical and experimental data available in literature.

Ilamparuthi and Muthukrishnaiah (1999) investigated anchors of very large uplift capacities that are required to support offshore structures at

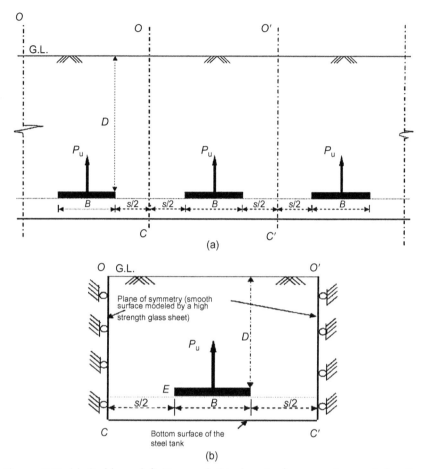

Figure 3.34 (a) Problem definition and (b) domain for carry in grout for the experiments by Kumar and Kouzer (2008).

great water depths as illustrated in Figs. 3.36 and 3.37. The capacities of plate and mushroom type anchors are generally estimated based on the shape of rupture surface. An attempt has been made to delineate the rupture surfaces of anchors embedded in submerged and dry sand bed sat at various depths. The results exhibited two different modes of failure depending on the embedment ratio, namely, shallow and deep anchor behavior. The load−displacement curves exhibited three- and two-phase behaviors for shallow and deep anchors, respectively. Negative pore water pressures recorded in submerged sand also exhibited variation similar to that of pullout load versus anchor displacement.

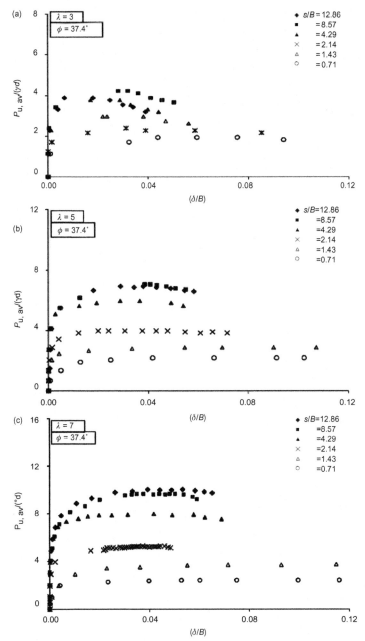

Figure 3.35 The variation pullout ultimate capacity for different values of (s/B) for Ø = 37.4 at (a) λ = 3, (b) λ = 5, and (c) λ = 7.

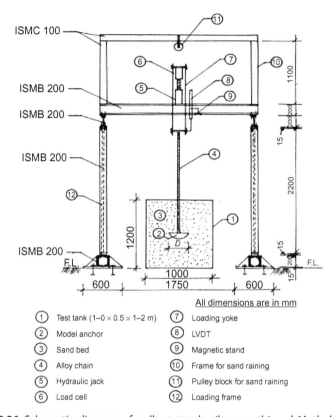

ISMC 100

ISMB 200

ISMB 200

ISMB 200

ISMB 200

All dimensions are in mm

①	Test tank (1–0 × 0.5 × 1–2 m)	⑦	Loading yoke
②	Model anchor	⑧	LVDT
③	Sand bed	⑨	Magnetic stand
④	Alloy chain	⑩	Frame for sand raining
⑤	Hydraulic jack	⑪	Pulley block for sand raining
⑥	Load cell	⑫	Loading frame

Figure 3.36 Schematic diagram of pullout test by Ilamparuthi and Muthukrishnaiah (1999).

Figure 3.37 Failure pattern of anchor plate in dense sand by Ilamparuthi and Muthukrishnaiah (1999).

From the experimental investigations carried out on half-cut models of flat circular and curved anchors embedded at different depths in dry and submerged sand beds of different densities, the following conclusions are arrived at:

1. Two types of rupture surface were observed, depending on the embedment ratio, one emerging to the surface of the sand bed and the other confined within the sand bed, irrespective of shape of anchor, density of sand bed and dry or submerged condition. A transition exists between the two types of failure, giving rise to the concept of "critical embedment ratio."

2. For embedment ratios less than the critical embedment ratio (shallow anchors), the rupture surface is a gentle curve, convex upwards, which can be closely approximated to a plane surface. The plane surface makes an angle of $\emptyset/2 \pm 2°$ with the vertical irrespective of density, submergence of sand bed and shape of anchor.

3. For embedment ratios greater than critical embedment ratio (deep anchors), the rupture surface emerging from the edge of the anchor makes an angle of 0.8 with the vertical irrespective of the density of the sand bed and the rupture surface is confined within the sand bed.

4. The load versus displacement relationship is different for shallow and deep anchors. A three-phase behavior for shallow anchors and two-phase behavior for deep anchors were observed irrespective of density and submergence of the sand bed.

5. For the densities studied in the submerged sand bed, the measured pore water pressures were negative. The variation of negative pore water pressure with anchor displacement is similar to that of pullout load versus anchor displacement.

Liu et al. (2010a,b) reported a test on cohesionless soil displacement field during anchor uplifting as illustrated in Fig. 3.38. The cohesionless soil displacement is calculated using the digital image correlation (DIC) method from two images: one is taken at the initial stage and the other is at the peak pullout load moment. The failure pattern is identified by locating the maximum shear strains deduced from the dense conditions: (1) in loose sand, the shearing bands start at the edges of the anchor plate an coverage and from a bell shape above the anchor; (2) while in dense sand these shear bands extend outward to the ground surface with an inclination angle with the vertical of approximately 1/2 to 1/3 the friction angle of soil.

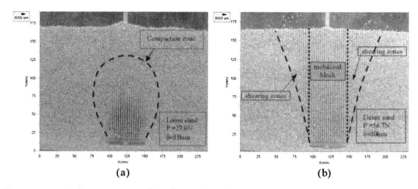

Figure 3.38 Failure pattern of anchor plates by Liu et al. (2010): (a) loose sand and (b) dense sand.

3.5 UPLIFT CAPACITY FOR REINFORCED ANCHOR PLATES

Symmetrical anchor plates are a foundation system that can resist tensile load with the support of the soil around which the anchor plate is embedded. It is used by soil structures as a structural member, primarily to resist uplift loads and overturning moments and to ensure the structural stability. A wide variety of soil anchor systems (plate, irregular shaped anchor, grouted and helical anchors) have been developed in order to satisfy the increase in foundations to resist the uplift responses. Engineers have shown that the uplift response can be improved by grouping the symmetrical anchor plates, increasing the unit weight, embedment ratio and the size of symmetrical anchor plates. The use of geosynthetics has been found to provide a possible solution in symmetrical anchor plate responses. Research into the uplift response of symmetrical anchor plates embedded in nonreinforced soil has been reported by Balla (1961), Meyerhof and Adams (1968), Vesic (1971, 1972), Hanna et al. (1972), Meyerhof (1973), Neely et al. (1973); Baset (1977), Das (1978, 1980), Rowe and Davis (1982), Saran et al. (1986), and Dickin (1987), although research in the area of symmetrical anchor plates embedded in reinforced soil such as Johnston (1984), Subbarao et al. (1988), Singh (1992), Rajagopal and SriHari (1996), Niroumand and Kassim (2010) were less extensive. The use of a symmetrical anchor plate system was first documented by Subbarao et al. (1988). Selvadurai (1989) investigated the performance of geogrids for anchoring 150 mm diameter and 850 mm long pipelines embedded in sand. Andreadis and Harvey (1981), Ghaly et al. (1991), Krishnaswamy and Parashar (1992),

Ilamaparuthi and Muthukrishnaiah (2001) and Niroumand (2011) were the few researchers that worked with reinforced sand and an uplift test. Krishnaswamy and Parashar (1992) carried out the experimental test in both reinforced sand beds and other nonreinforced beds. A more extensive experimental study was carried out by Selvadurai (1993) to evaluate the treatment of a 215-mm-diameter pipe with a length of 1610 mm embedded in reinforced sand beds. The inclusion of geogrids immediately above the pipeline in an inclined setup increased the uplift response by about 80%. This increased load was sustained for uplift displacements, and extended the ductility.

The uplift response of symmetrical anchor plates embedded in sand bed with geosynthetic reinforcement material was studied by Krishnaswamy and Parashar (1994). Krishnaswamy and Parashar (1994) investigated the uplift response of symmetrical anchor plates such as circular anchor plates (60 mm in diameter) and rectangular plates (53 mm wide, with lengths varying from 23.8 to 53 mm) embedded in clay, and sand with and without geosynthetics. Placing the geosynthetics directly on the symmetrical anchor plate was proved to be beneficial in achieving maximum increase in the uplift response although they found that two layers of geogrid reinforcement did not hugely increase the uplift capacity. Ilamparuthi and Dickin (2001a,b) undertook an extensive study on the treatment of belled pile anchors in reinforced sand beds and formulated a hyperbolic method for calculating the breakout factor. Ravichandran and Ilamparuthi (2004) evaluated the treatment of rectangular anchor plates in nonreinforced and reinforced cohesionless soil beds. Kingshri et al. (2005) evaluated two series of experimental tests to understand the influence of stiffness and opening size of geosynthetic reinforcement materials on the uplift capacity of rectangular anchor plates. The first series of tests used a combination of geocomposite and geogrid materials, and the second series of tests were on two layers of geogrid as reinforcements. It was concluded that the performance of a geocomposite and geogrid (two layer) combination was found to be more effective in resisting uplift response than two combined layers of geogrids.

The performance of symmetrical anchor plates has been investigated by many researchers in nonreinforced sand, but relatively little is known about the performance of symmetrical anchor plates in reinforced soil beds. Johnston (1984) investigated the pullout response of geogrids. Subbarao et al. (1988) evaluated the improvement in pullout load by using geotextiles as ties to symmetrical anchor plates embedded in sand.

Experimental tests were conducted on reinforced concrete model symmetrical anchor plates of cylindrical and belled shape, using polypropylene ties of width 55 mm and thickness 0.72 mm. Results showed that symmetrical anchor plates with geotextile ties offered greater uplift response than those without ties. Furthermore, the single layers of ties close to the symmetrical anchor plates were reported to be more effective than the use of multiple layers.

Increasing anchors' usefulness to resist and sustain uplift forces may be achieved by increasing the size and depth of an anchor or the improvement of soil in which these anchors are embedded, or both. In restricted situations, increasing the size and depth of an anchor may not be economical compared with other alternatives. On the other hand, soil improvement can be attained by the inclusion of soil reinforcement to resist larger uplift forces. However, very few investigations of the behavior of horizontal plates in a reinforced soil bed under uplift loads have been reported. Selvadurai (1989, 1993) reported significant enhancement, of the order of 80−100%, in the uplift capacity of pipelines embedded in fine and coarse-grained soil beds reinforced by inclusion of geogrids immediately above the pipeline in an inclined configuration. Krishnaswamy and Parashar (1994) studied the uplift behavior of circular plates and rectangular plates embedded in cohesive and cohesionless soils with and without geosynthetic reinforcement, and reported that the geo-composite reinforcement offered higher uplift resistance than both geo-grid and geotextile reinforcement. Ilamparuthi and Dickin (2001a,b) investigated the behavior of soil reinforcement on the uplift response of piles embedded in sand through model tests with cylindrical gravel-filled geogrid soil placed near the pile base as illustrated in Figs. 3.39 and 3.40. Authors reported increases in the uplift response of piles with many factors such as the diameter of the geogrid cell, sand density, pile bell diameter, and embedment.

El Sawwaf (2007) conducted a laboratory experimental investigation on strip anchor plates to investigate their uplift response in sand. Uplift response of symmetrical anchor plates located close to sandy slopes with and without geosynthetic reinforcement has been evaluated in tests. Strip anchor plates were used in the experimental work to research the effect of soil reinforcement on the uplift behavior of anchor plates using plane strain. Strip anchor plates made of steel of 498 mm in length, 6.0 mm in thickness, and 80 mm in width were made with a special hole 3.0 mm in diameter in the center and used in the research. The authors concluded

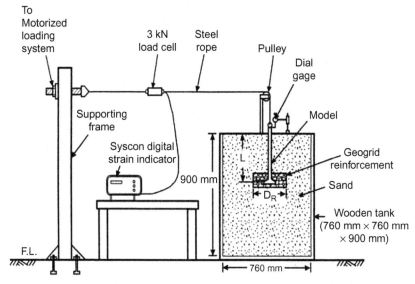

Figure 3.39 Schematic diagram of experimental test by Ilamparuthi and Dickin (2001a,b).

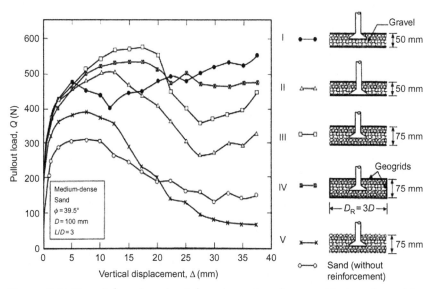

Figure 3.40 Nonreinforced and reinforced results of symmetrical anchor plates under uplift test by Ilamparuthi and Dickin (2001a,b).

that an increase in the ultimate pullout response of an anchor plate embedded to the slope crest and anchor plate improvement is very dependent on geosynthetic layer length, and increases significantly until the amount of beyond that further increase in the layer length does not show a significant contribution in the anchor resistance. Geosynthetic layers were placed to reinforce the slope as shown in Fig. 3.41.

Ilamparuthi et al. (2008) undertook two investigations on the submerged sand effect of symmetrical anchor plates on uplift response of treated nonreinforced and reinforced sand as illustrated in Figs. 3.42 and 3.43. Influence factors such as embedment ratio, density and number of geogrid layers were investigated. The authors reported that the load−displacement

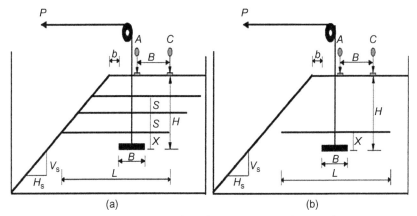

Figure 3.41 Geometric parameters of reinforced slope by El Sawwaf (2007).

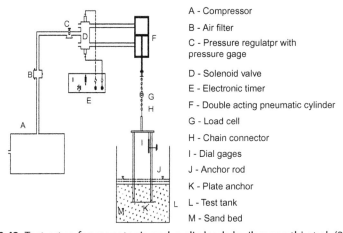

A - Compressor
B - Air filter
C - Pressure regulatpr with pressure gage
D - Solenoid valve
E - Electronic timer
F - Double acting pneumatic cylinder
G - Load cell
H - Chain connector
I - Dial gages
J - Anchor rod
K - Plate anchor
L - Test tank
M - Sand bed

Figure 3.42 Test setup for monotonic and cyclic loads by Ilamparuthi et al. (2008).

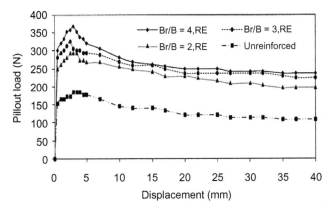

Figure 3.43 Pullout response on symmetrical anchor plates in medium dense sand bed for $H/B = 4$ by Ilamparuthi et al. (2008).

behavior of nonreinforced and reinforced sand for a given density and embedment ratio were similar except for higher peak and residual loads due to the geogrid materials. In the case of a cyclic load, the displacement of symmetrical anchor plate was increased with decreasing rate and reached almost a constant value after 350 cycles.

Niroumand et al. (2013) evaluated the uplift response of horizontal anchor plates with and without geogrid and GFR reinforcement in model tests and numerical simulations. Many items of reinforcement layers were used to reinforce the sandy soil over a symmetrical anchor plate. Different factors such as the relative density of sand, embedment ratios along with geogrid and GFR parameters including size, number of layers, and the proximity of the layer to the anchor plate have been investigated in a scale model. The failure mechanism and the associated rupture surface were observed and evaluated. GFR is a new tied up system with innovative design, made from fiber reinforcement polymer (FRP) which anchors the geogrid into the soil. Test results showed that using GFR reinforcement has a significant effect in improving the uplift capacity of circular anchor plates. It was found that inclusion of one layer of GFR that is located resting directly on top of the anchor plate was more effective in enhancing the anchor capacity than reinforcing the geogrid itself. It was found that inclusion of one layer of geogrid on the symmetrical anchor plate improved the uplift capacity by 19% as compared to the same symmetrical anchor plate embedded without reinforcement. However with the inclusion of GFR the uplift response improved further to 29%. The single layer geogrid was also more effective in enhancing the uplift capacity compared to the multilayer geogrid reinforcement. On the other hand, the single layer geogrid with GFR gave higher uplift capacity as compared to single layer geogrid. This is due to the additional anchorage provided by the GFR at each level of reinforcement. In general the results show that the uplift capacity of symmetrical anchor plates in loose and dense sand can be significantly increased by the inclusion of geogrid with GFR. It was also observed that the inclusion of geogrid with GFR reduces the requirement for higher L/D ratio to achieve a required uplift capacity. The laboratory and numerical analysis results are found to be in agreement in terms of breakout factor and failure mechanism pattern.

3.6 HORIZONTAL ANCHOR PLATE GROUPINGS

The total pullout net ultimate capacity of group anchor plates is weaker than a single anchor plate..The reason is related to center-to-center spacing of the horizontal anchor plates which is small, under a pullout test although the failure zone will be changed. Researchers, such as Das and Jin-Kaum (1987), Murray and Geddes (1996), Meyerhof and Adams (1968), have investigated the use of group anchor plates. Here, we will focus on the Meyerhof and Adams method because sometimes engineers need to design group anchor plates in geotechnical projects. In their experiment, Meyerhof and Adams assumed the circular shapes with equal center-to-center spacing of the anchor plates (s), number of rows (m), number of columns (n), and gross ultimate uplift capacity of anchor group, $Q_{ug(g)}$. They created the below function:

$$Q_{ug(g)} = Q_{ug} + W_g$$

where Q_{ug} is the net ultimate uplift capacity of the anchor group and W_g is effective self-weight of anchor plates and shafts.

Meyerhof and Adams's method suggested the net ultimate capacity of horizontal anchor plates as

$$Q_{ug} = \gamma L^2 \left(a + b + S_f \left(\frac{\pi}{2} \right) D \right) K_u \tan \varphi + w_s$$

where S_f is shape factor $(1 + m(L/D))$

K_u, nominal uplift coefficient

w_s, effective weight of the sand placed the horizontal anchor plate

$a = s(n-1)$

$b = s(m-1)$.

Meyerhof and Adams's method could be defined group efficiency, η, as

$$\eta = \frac{Q_{ug}}{mnQ_u}$$

3.7 DESIGN CONSIDERATION

The importance of horizontal anchor plates are clear in geotechnical projects. They can be used in various projects such as transmission towers, tents, pipelines, and related projects. A recommended $2-2.5$ factor of safety should be used in the analysis and design of uplift capacity of horizontal anchor plates. Most published methods provide for axisymmetric projects, although Meyerhof and Adams (1968) provided another set of plate anchors that could be used in conjunction with rectangular plates. For deep anchor plates, Meyerhof and Adams's method is helpful as most other methods are designated for shallow conditions ($L/D \leq 5$). Based on Meyerhof and Adams's method the changeable ratio of strip plates to square plates can be defined as

$$\frac{(L/D)_{cr-Strip}}{(L/D)_{cr-Square}} = 1.5$$

Generally two methods could be used in determining the allowable net ultimate capacity of horizontal anchor plates; the first method used a factor of safety, F_s, in the equation:

$$Q_u(\text{allowable}) = \frac{Q_u}{F_s}$$

The second method could be used in determining the load−displacement relationship which corresponds to predetermined allowable displacement of horizontal anchor plate.

If horizontal anchor plates are placed under repeated loads then these loads established cyclic loads that could create uplift loads. The current knowledge on this subject is so limited, with the main information deriving from Andreadis et al. (1981) who evaluated the performance of circular anchor plates in dense saturated sand under cycling loads.

3.8 CONCLUSION

It is clear that much research has been undertaken of the performance of horizontal anchor plates in the sand. This research has included using different horizontal anchor plates and soil parameters. Inevitably such a wide range of parameters will contribute to conflicting conclusions for the ultimate pullout load of the horizontal anchor plates. Some of the works did not include measurements for the internal friction angle, anchor roughness, and anchor size. However, most researchers obtained their internal friction angle using the direct shear test or tri-axial compression test. Unfortunately, the results obtained from the laboratory tests are typically a specific problem and are difficult to extend and develop to field problems due to the different materials or the geometric parameters in the field scale. Moreover, geotechnical engineers are lacking in reported experimental data, and this makes the comparison between the theory and numerical data very difficult.

Most soil anchor research has been concerned with the uplift problem on embedded in nonreinforced soils under pullout test. Symmetrical anchor plates are a foundation system that can resist tensile load with the support of surrounding soil in which a symmetrical anchor plate is embedded. Engineers and researchers proved that the uplift response can be improved by grouping the symmetrical anchor plates, increasing the unit weight, embedment ratio, and the size of symmetrical anchor plates. This chapter discussed the definitions of horizontal anchor plates in cohesionless soil. Horizontal anchor plates have their own innate limitations, such as the need to excavate and backfill in various geotechnical projects, however they have great importance in most geotechnical projects. Innovation of geosynthetics and GFR in the field of geotechnical engineering as reinforcement materials are one possible solution in symmetrical anchor plate responses. Unfortunately the importance of reinforcement in submergence, cyclic response of anchor plates, and the pullout response of group anchor plates have received very little attention by researchers and engineers.

REFERENCES

Andreadis, A., Harvey, R., Burley, E., 1981. Embedded anchor response to uplift loading. J. Geotech. Eng 107 (1), 59–78.
Baker, W.H., Konder, R.L., 1966. Pullout load capacity of a circular earth anchor buried in sand. Highw. Res. Rec. 108, 1–10.

Balla, A. 1961. The resistance of breaking-out of mushroom foundations for pylons. Proceedings, 5th International Conference on Soil Mechanics and Foundation Engineering, Vol. 1, Paris, pp. 569−576.

Basudhar, P.K., Singh, D.N., 1994. A generalized procedure for predicting optimal lower bound break-out factors of strip anchors. Geotechnique 44 (2), 307−318.

Baset, R.H., 1977. Underreamed ground anchors. Proc. Annu. Conf. Res. Med. Educ 1, 11−17.

Caquot, A.I., Kerisel, J., 1948. Tables for the calculation of passive pressure, active pressure, and bearing capacity of foundations." Libraire du Bureau des Longitudes, de L'ecole Polytechnique. Imprimeur-Editeur, Paris Gauthier- villars, 120.

Clemence, S.P., Veesaert, C.J. 1977. Dynamic pullout resistance of anchors in sand. Int Symp on Soil Struct Interaction. Roorkee, India; 3 January 1977 through 7 January 1977, pp. 389−397.

Das, B.M., 1978. Model tests for uplift capacity of foundations in clay. Soils Found. 18 (2), 17−24.

Das, B.M., 1980. A procedure for estimation of ultimate uplift capacity of foundations in clay. Soils Found. 20 (1), 77−82.

Das, B.M., 1990. Earth Anchors. Elsevier, Amsterdam.

Das, B.M., Seeley, G.R., 1975a. Breakout resistance of shallow horizontal anchors. J. Geotech. Eng., ASCE 101 (9), 999−1003.

Das, B.M., Seeley, G.R., 1975b. Load displacement relationship for vertical anchor plates. J. Geotech. Eng., ASCE 101 (7), 711−715.

Dickin, E.A., 1987. Uplift Behaviour of Horizontal Anchor Plates in Sand. J. Mech. Found. Eng. Div 114 (SM11), 1300−1317.

Dickin, E.A., 1988. Uplift behaviour of horizontal anchor plates in sand. J. Geotech. Eng 114 (11), 1300−1317.

Dickin, E.A., Laman, M., 2007. Uplift response of strip anchors in cohesionless soil. J. Adv. Eng. Softw. 1 (38), 618−625.

Downs, D.I., Chieurzzi, R., 1966. Transmission tower foundations. J. Pow. Div. ASCE 88 (2), 91−114.

El Sawwaf, M.A., 2007. Uplift behavior of horizontal anchor plates buried in geosynthetic reinforced slopes. Geotech. Testing J. 30 (5), 418−426.

Fargic, L., Marovic, P., 2003. Pullout capacity of spatial anchors. J. Eng. Comput. 21 (6), 598−700.

Frydman, S., Shamam, I., 1989. Pullout capacity of slab anchors in sand. Can. Geotech. J. 26, 385−400.

Ghaly, A.M., 1977. Soil restraint against oblique motion of pipelines in sand. Discuss. Can. Geotech. J. 34 (1), 156−157.

Ghaly, A.M., Hanna, A.M., 1994a. Model investigation of the performance of single anchors and groups of anchors. Can. Geotech. J 31 (2), 273−284.

Ghaly, A.M., Hanna, A.M., 1994b. Ultimate pullout resistance of single vertical anchors. Can. Geotech. J. 31 (5), 661−672.

Ghaly, A.M., Hanna, A.M., Hanna, M., 1991. Uplift behavior of screw anchors in sand. I: dry sand. J. Geotech. Eng. ASCE 117 (5), 773−793.

Ilamparuthi, K., Muthukrisnaiah, K., 1999. Anchors in sand bed: delineation of rupture surface. Ocean Eng 26, 1249−1273.

Ilamparuthi, K., Muthukrishnaiah, K. 2001. "Breakout capacity of seabed anchors due to snap loading." Proceedings of International Conference in Ocean Engineering, IIT Madras, Chennai, India, pp. 393−400.

Ilamparuthi, K., Dickin, E.A., 2001a. Predictions of the uplift response of model belled piles in geogrid-cell-reinforced sand. Geotext Geomembrane 19, 89−109.

Ilamparuthi, K., Dickin, E.A., 2001b. The influence of soil reinforcement on the uplift behaviour of belled piles embedded in sand bed. Geotext Geomembrane 19, 1−22.

Ilamparuthi, K., Ravichandran, P., Mohammed Toufeeq, M., 2008. Study on uplift behavior of plate anchor in Geogrid reinforced sand bed. Geotech Spec. Publ.(181).

Johnston, R.S., 1984. Pull-Out Testing of Tensar Geogrids, Master's Thesis. University of California, Davis, California, USA, 179p.

Kananyan, A.S., 1966. Experimental investigation of the stability of bases of anchor foundations. Osnovanlya, Fundamenty i mekhanik Gruntov 4 (6), 387−392.

Kingshri, A., Ilamparuthi, K., Ravichandran, P.T. 2005. Enhancement of uplift capacity of anchors with Geocomposite. Proceeding of National Symposium on Geotechnical prediction methods, "Geopredict 2005", IIT Madras, Chennai, pp. 148−152.

Koutsabeloulis, N.C., Griffiths, D.V., 1989. Numerical modelling of the trap door problem. Geotechnique 39 (1), 77−89.

Krishnaswamy, N.R., Parashar, S.P., 1992. Effect of Submergence on the uplift resistance of footings with geosynthetic inclusion. Proc. Indian. geotechnic. conf., Surat., India 333−336.

Krishnaswamy, N.R., Parashar, S.P., 1994. Uplift behaviour of plate anchors with Geosynthetics. Geotext. Geomembrane 13, 67−89.

Kumar, J., Kouzer, K.M., 2008. Vertical uplift capacity of horizontal anchors using upper bound limit analysis and finite elements. Can. Geotech. J. 45, 698−704.

Kumar, J., Bhoi, M.K., 2008. Interference of multiple strip footings on sand using small scale model tests. Geotech. Geol. Eng. 26 (4), 469−477.

Kuzer, K.M., Kumar, J., 2009. Vertical uplift capacity of two interfering horizontal anchors in sand using an upper bound limit analysis. J. Comp. Geotech. 1 (36), 1084−1089.

Lade, P.V., Duncan, J.M., 1975. Elasto-plastic stress−strain theory for cohesionless soil. J. Soil Mech. Found. Div. ASCE 101 (10), 1037−1053.

Liu, M., Liu, J., Gao, H. 2010a. Displacement field of an uplifting anchor in sand. Geotechnical Special Publication, Issu 205 GSP, 261−266.

Liu, H., Li, Y., Yang, H., Zhang, W., Liu, C., 2010b. Analytical study on the ultimate embedment depth of drag anchors. Ocean Eng. 37 (14−15), 1292−1306.

Mariupolskii, L.G., 1965. The bearing capacity of anchor foundations. SMFE, Osnovanlya, Fundamenty i mekhanik Gruntov 3 (1), 14−18.

Meyerhof, G.G., Adams, J.I., 1968. The ultimate uplift capacity of foundations. Can. Geotech. J. 5 (4), 225−244.

Meyerhof, G.G., 1951. Ultimate bearing capacity of footings on sand layer overlaying clay. Can. Geotech. J 11 (2), 223−229.

Meyerhof, G.G., 1973. Uplift Resistance of Inclined Anchors and Piles. Int. Conf. Mec. Found. Eng. 21 (1), 167−172.

Merifield, R., Sloan, S.W., 2006. The ultimate pullout capacity of anchors in frictional soils. Can. Geotech. J 43 (8), 852−868.

Mors, H., 1959. The behaviour of most foundations subjected to tensile forces. Bautechnik 36 (10), 367−378.

Murray, E.J., Geddes, J.D., 1987. Uplift of anchor plates in sand. J. Geotech. Eng., ASCE 113 (3), 202−215.

Murray, E.J., Geddes, J.D., 1989. Resistance of passive inclined anchors in cohesionless medium. Géotechnique 39 (3), 417−431.

Murray, E.J., Geddes, J.D., 1996. Plate anchor groups pulled vertically in sand. J. Geotech. Eng 122 (7), 509−516.

Neely, W.J., Stuart, J.G., Graham, J., 1973. Failure loads of vertical anchor plates in sand. ASCE J. Soil Mech. Found. Div. 99 (SM9), 669−685.

Niroumand, H., Kassim, K.A., 2010. Analytical and numerical studies of vertical anchor plates in cohesion-less soils. EJGE 15 (L), 1140−1150.

Niroumand, H., Kassim, K.A., Nazir, R., 2013. The influence of soil reinforcement on the uplift response of symmetrical anchor plate embedded in sand. Measurement 46 (8), 2608−2629.

Pearce, A., 2000. Experimental investigation into the pullout capacity of plate anchors in sand." MSc thesis. University of Newcastle, Australia.

Rajagopal, K. and SriHari, V., 1996, "Analysis of Anchored Retaining Walls", *Earth Reinforcement*, Ochiai, H.,Yaufuku, N. and Omine, K., Editors, Balkema, 1997, Vol.1, Proceedings of the International Symposium on Earth Reinforcement (1996), Fukuoka, Kyushu, Japan, 475-479.

Ramesh Babu, R., 1998. Uplift Capacity and Behaviour of Shallow Horizontal Anchors in Soil, Ph.D., Thesis. Dept. of Civil Engg., Indian Institute of Science, Bangalore.

Rowe, R.K., Davis, E.H., 1982. The behaviour of anchor plates in sand. Geotechnique 32 (1), 25−41.

Rowe, R.K., 1978. Soil structure interaction analysis and its application to the prediction of anchor behaviour. PhD thesis. University of Sydney, NSW, Australia.

Saeedy, H.S., 1987. Stability of circular vertical anchors. Can. Geotech. J 24, 452−456.

Sakai, T., Tanaka, T., 1998. Scale effect of a shallow circular anchor in dense sand. Soils and Found., Japan 38 (2), 93−99.

Saran, S., Ranjan, G., Nene, A.S., 1986. Soil anchors and constitutive laws. J. Geotech. Eng 112 (12), 1084−1100.

Sarac, D.Z., 1989. Uplift capacity of shallow buried anchor slabs. Proc., 12th Int. Con. Mech. Foun. Eng 12 (2), 1213−1218.

Selvadurai, A.P.S., 1989. The enhancement of the uplift capacity of buried pipelines by the use of geogrids. J. Geotech. Test. ASTM 12, 211−216.

Selvadurai, A.P.S., 1993. Uplift behaviour of strata grid anchored pipelines embedded in granular soils. Geotech. Eng 24, 39−55.

Sergeev, I.T., Savchenko, F.M., 1972. Experimental investigations of soil pressure on the surface of an anchor plate. Soil Mech. Found. Eng. 9 (5), 298−300.

Singh, R.B., 1992, "Anchored Earth Technique using Semi-Z shaped Mild Steel Anchors." *Earth Reinforcement Practice*, Ochiai, H., Hayashi, S. and Otani, K., Editors, Balkema, 1992, Vol. 1, Proceedings of the International Symposium on Earth Reinforcement, Fukuoka, Kyushu, Japan, 1992, 419−424.

Smith, C.C., 1998. Limit loads for an anchor/trapdoor embedded in an associated cou-lomb soil. Int. J. Numer. Anal. Methods Geomech 22 (11), 855−865.

Sloan, S.W., 1988. Lower bound limit analysis using finite elements and linear program-ming. Int. J. Numer. Anal. Methods Geomech 12 (1), 61−67.

Subbarao, C., Mukhopadhyay, S., Sinha, J. 1988, "Geotextile Ties to Improve Uplift Resistance of Anchors." Proceedings of the 1st Indian Geotextile Conference on Reinforced Soil and Geotextiles, F3−F8.

Sutherland, H.B., 1965. Model studies for shaft raising through cohesionless soils. Proc., 6th. Int. Conf. Mech. Found. Eng 2, 410−413.

Tagaya, K., Scott, R.F., Aboshi, H., 1988. Pullout resistance of buried anchor in sand. Soils Found. Japan 28 (3), 114−130.

Tagaya, K., Tanaka, A., Aboshi, H., 1983. Application of finite element method to pull-out resistance of buried anchor. Soils and Found. Japan 23 (3), 91−104.

Vermeer, P.A., Sutjiadi, W., 1985. The uplift resistance of shallow embedded anchors. Proc., 11th Int. Conf. Mech. Found. Eng., San Francisco 4, 1635−1638.

Vesic, A.S., 1971. Breakout resistance of objects embedded in ocean bottom. J. Mech. Found. Div., ASCE 97 (9), 1183−1205.

Vesic, A.S., 1972. Expansion of cavities in infinite soil mass. J. Mech. Found. Div., ASCE 98 (3), 265−290.

CHAPTER 4

Horizontal Anchor Plates in Cohesive Soil

4.1 INTRODUCTION

Research into the uplift capacity of soil anchor plates in cohesive soils is fairly sparse, when compared to that of cohesionless soil, particularly when it comes to clay. Engineers need to have suitable methods for design and construction of soil anchor plates in clay because most soil includes a clay layer or a layer of C-Ø soil. Different types of soil anchor plates, such as square, circular, strip, and rectangular plates can be embedded into clay soil. Clay soils differ greatly from sand soils as there is an undrained cohesion of clay, C_u. When the soil anchor plate is located at a suitable embedment depth, under an uplift test, the soil above the plate will be compressed while the soil below the plate receives stresses, increasing the pore water pressure above the plate and decreasing the pore water pressure below the plate. As illustrated in Fig. 4.1, the uplift capacity of soil anchor plates in clay is:

$$Q_{u(g)} = Q_u + w_a + U$$

where $Q_{u(g)}$ is the gross ultimate uplift capacity,

Q_u = net ultimate uplift capacity

w_a = effective weight of the anchor

U = suction force in below of anchor plate that is assumed as neglected because the information is limited on it at this moment.

Thus Q_u is equal to

$$Q_{u(g)} - w_a = Q_u$$

This chapter discusses some theories that can assist any analysis of the uplift capacity of soil anchor plates.

4.2 FAILURE PATTERN

Rowe and Davis (1982) suggested two categories for failure mechanism of soil anchor plates in clay: "immediate breakaway" and "no breakaway".

Design and Construction of Soil Anchor Plates.
DOI: http://dx.doi.org/10.1016/B978-0-12-420115-6.00004-7

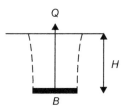

Figure 4.1 Horizontal anchor plate in clay.

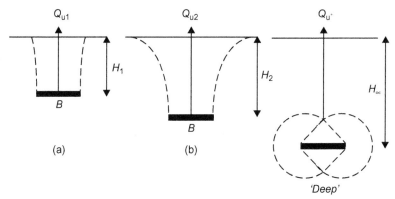

Figure 4.2 Types of failure mechanism of soil anchor plates by Merifield et al. (2003).

In the case of "immediate breakaway" the soil anchor interface cannot sustain tension; whilst in case of "no breakaway," the soil anchor interface sustains adequate tension. Merifield et al. (2003) classify the failure modes according to shallow and deep conditions (see Fig. 4.2).

Generally anchor plates are separated into shallow and deep conditions based on their effect on soil surface. In the shallow conditions, the failure mode affects the soil surface, whilst deep conditions do not affect the soil surface.

4.3 UPLIFT CAPACITY FOR ANCHOR PLATES IN CLAY

Most researchers in this area, such as Meyerhof and Adams (1968), Vesic (1971), Meyerhof (1973), Das (1978), Rowe and Davis (1982), Saran et al. (1986), Rao and Kumar (1994), Merifield et al. (2001), Mehryar et al. (2002), Thorne et al. (2004), Song et al. (2008), Khatri and

Kumar (2009), and Wang et al. (2010) focused on analysis and prediction of uplift capacity of soil anchor plates in clay. Some of this was undertaken by experimentation; otherwise some were analyzed anchor plates using numerical methods. A discussion of the research now follows.

4.3.1 Meyerhof's Method

Meyerhof and Adams (1968) and Meyerhof (1973) suggested related equations for ultimate uplift capacity and breakout factors in various types of soil anchor plates in clay. Meyerhof (1973) investigated a number of model uplift tests in clay. He suggested the equation:

$$Q_u = A(\gamma L + N_q C_u)$$

where N_q for square plates

$$N_q = 1.2(L/D) \leq 9$$

For circular plate

$$N_q = 1.2(L/D) \leq 9$$

For strip plate

$$N_q = 0.6(L/D) \leq 8$$

where $(L/D)cr = 9/1.2 = 7.5$
and $(L/D)cr = 8/0.6 = 13.5$

Meyerhof and Adams (1968) and Meyerhof (1973) investigated the uplift capacity and breakout factors of horizontal circular anchor plates in clay. Meyerhof and Adams (1968) suggested a semitheoretical function for anchor plates (rectangular, circular and strip plates) in C-Ø soils, although they were first authors showed the shape factor in practice. The failure zone was very difficult in this case because tensile cracks were not clear, although Meyerhof (1973) continued his research on ultimate uplift capacity of anchor plates in clay. He proposed related functions on circular, square, and strip plates in stiff clay. He defined the importance uplift capacity in undrained condition rather than drained condition in clay because, with time, the negative pore water pressures increased, thus

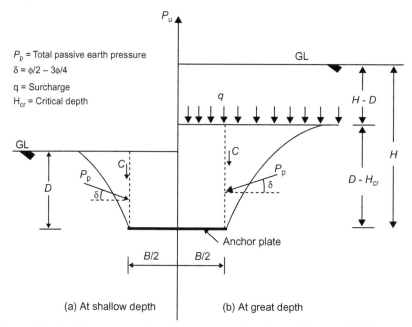

Figure 4.3 The failure mode of anchor plate by Meyerhof and Adams (1968).

Table 4.1 Fc in clay for strip and circular plates

Anchor shape			L/D		
	0.5	1	1.5	2	2.5
Circular	1.76	3.8	6.12	11.6	30.3
Strip	0.81	1.61	2.42	4.04	8.07

softening the soil. Fig. 4.3 illustrates the failure zone of anchor plate by Meyerhof and Adams (1968).

4.3.2 Vesic's Method

Vesic (1971) investigated an expanding spherical cavity near ground surface in soil. He suggested the below function for clay:

$$Q_u = A(\gamma LF_q + CNF_c)$$

where F_c is breakout factor based on Table 4.1
and C is soil cohesion

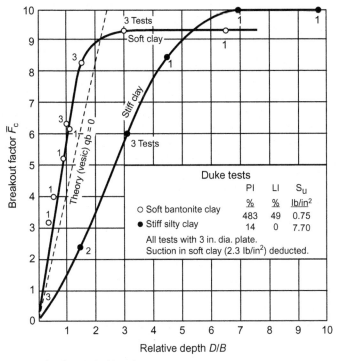

Figure 4.4 F_c in Clay by Vesic (1971).

If it is an undrained condition considered, then $\varnothing = 0$ and C_u proposed is thus:

$$Q_u = A(\gamma L + C_u F_c)$$

Looking at Fig. 4.4 we can see that the breakout factor increased up to maximum amount then becomes constant; this maximum is F_c^* and showed the embedment ratio in critical conditions, (L/D)cr. Vesic's method separated anchor plates based on embedment ratio in critical limitation to shallow and deep conditions as:

If $(L/D) \leq (L/D)$cr for shallow anchor plates

and if $(L/D) \geq (L/D)$cr for deep anchor plates

This information means that Vesic's method could be used for the analysis and design of anchor plates in clay.

4.3.3 Das's Method

Das (1978) investigated the uplift response of circular anchor plates in the laboratory. He used undrained cohesion, C_u, around $5.18-172.5$ kN/m².

Das's method proposed the breakout factor for shallow anchor plates based on below function:

$$N_q = n(L/D) \leq 8 - 9$$

where n = constant amount between 2 and 5.9, although n is a function of C_u.

He suggested various functions based on square and rectangular anchor plates with width of 50.8 mm and he proposed the below function:

$$(L/D)cr\text{-}S = 0.107 \, C_u + 2.5 \leq 7$$

where $(L/D)cr\text{-}S$ is critical embedment ratio of square anchor plates, although this function is similar to circular anchor plates in practice, and C_u is undrained cohesion (kN/m^2)

For rectangular anchor plates, Das's method suggested the below function:

$$(L/D)cr\text{-}R = (L/D)cr\text{-}S(0.71 + 0.27(B/D)) \leq 1.55(L/D)cr\text{-}S$$

where $(L/D)cr\text{-}R$ is the critical embedment ratio of rectangular anchor plates.

Das (1978) investigated the uplift response of anchor plates in clay in practice but he continued his research in 1980. Das (1980) proposed new functions on shallow and deep anchor plates because he suggested new factors (α' and β'):

$$\alpha' = \frac{\left(\frac{L}{D}\right)}{\left(\frac{L}{D}\right)cr}$$

$$\beta' = \frac{F_c}{F_c^*}$$

$$F_{c-R}^* = 7.56 + 1.44(D/B)$$

where F_{c-R}^* is breakout factor for deep anchor rectangular anchor.

Das's method suggested the following steps for analyzing the uplift response of anchor plates in clay:

1. Identify the undrained cohesion, C_u
2. Identify the critical embedment ratio by $(L/D)cr\text{-}S = 0.107 \, C_u + 2.5 \leq 7$
3. Identify the embedment ratio

4. If $(L/D) \leq (L/D)cr$ then it is shallow anchor plates unless deep anchor plates.
5. If $(L/D) \leq (L/D)cr$ then $Q_u = A(\beta' F_c^* C_u + \gamma L)$
6. If $(L/D) \geq (L/D)cr$ then $F_c = F_c^*$ thus $Q_u = A([7.56 + 1.44(D/B)] C_u + \gamma L)$

Das's method could be used to predict the uplift response of anchor plates in clay.

4.3.4 Rowe and Davis's Method

Rowe and Davis (1982) investigated the undrained response of soil anchor plates using the finite element method in clay. They compared their results with existing experimental data. The main activity focused on performance of strip anchor plates under vertical and horizontal loads. Elasto-plastic and the finite element method was selected for analysis in their research because they suggested the use of the plastic zones of soil anchor plates. The main result of these researchers was defined on only shallow anchor plates with embedment ratio of less than 3. Figs. 4.5 and 4.6 illustrated the plastic zones of anchor plates in clay.

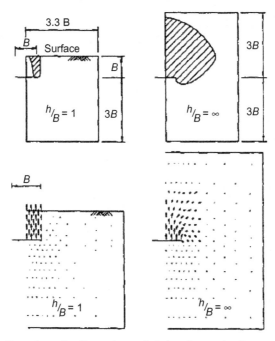

Figure 4.5 Plastic regions at collapse; immediate breakaway, by Rowe and Davis (1982).

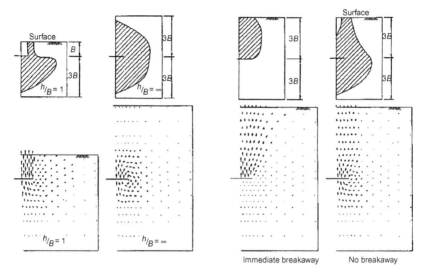

Figure 4.6 Plastic regions at failure; no breakaway, by Rowe and Davis (1982).

4.3.5 Saran's Method

Saran et al. (1986) predicted the load-displacement of shallow anchor plates in C-Ø soils. They used hyperbolic stress—strain curves in their analytical method. They analyzed all shapes of soil anchor plates in deep and shallow conditions although the size, anchor shape and soil factors were considered in the method.

4.3.6 Additional Related Research

Rao and Kumar (1994) suggested a method based on log-spiral failure mode that could be used for strip and square plates in various soils. Merifield et al. (2001) proposed the use of finite element method and limit analysis for evaluation of uplift response of soil anchor plates. They analyzed various parameters and suggested an equation for shallow and deep anchor plates based on dimensionless overburden ratio. They defined the shallow and deep differences based on overburden pressure because, up to a limiting value, the condition changed from shallow to deep, although also suggested a relationship for soil anchor plates under uplift loads between homogeneous and inhomogeneous soils.

The effect of suction in soil anchor plates was a topic that Mehryar et al. (2002) investigated. They considered nonattached and attached bases in soil anchor plates under uplift loading in finite element analysis. They suggested the importance of suction in soil anchor plates because they find that in a nonattached base, deep embedment limit is close to 4; but for an attached base, it is only 2.

Thorne et al. (2004) analyzed the importance of various factors in soil anchor plates. They analyzed H/B, $\gamma H/c$ & uc/c, where H is the embedment depth, B is the width of plate, γ is the unit weight of soil, c is the cohesion and uc is the maximum tensile force sustained by pore water of soil. These variables showed a close connection to the uplift capacity of soil anchor plates. A good result was related to the important separation of clay from the anchor plate in saturated clay if the effective stress below an anchor plate was zero.

Song et al. (2008) investigated the response of circular and strip anchor plates by finite element method in consolidated clays, although they did not consider suction below the anchor plates. Khatri and Kumar (2009) evaluated the importance of limit analysis and finite element method on uplift capacity of circular anchor plate. They suggested the importance of critical depth for increasing the breakout but after it the breakout was constant. The effect of roughness of anchor plates in strip, circular, and rectangular plates investigated by Wang et al. (2010). They found the minimal effect of roughness of anchor plates by finite element analysis although they suggested the overestimate the uplift capacity of anchor plates by lower bound limit analysis and small strain finite element analysis. Niroumand et al. (2012) focused on a systematic analysis review on soil anchor plates in clay.

4.4 GENERAL REMARKS

Most investigation methods identified that a deformation occurred before the ultimate collapse of a building. It is easy to see that that this factor is so important for design and that engineers need to consider safety when using anchor plates. In clay conditions, the engineer also has to consider the likelihood of breakaway or no breakaway when designing structures.

4.5 CONCLUSION

Much more research is needed on the uplift response of anchor plates when dealing with clay soils. It is possible that geosynthetics and grid-fixed reinforced in the field of geotechnical engineering to be used as reinforcement materials may be a possible solution in symmetrical anchor plate responses. Unfortunately the importance of reinforcement in submergence, cyclic response of anchor plates, and the pullout response of group anchor plates have received very little attention by research and engineers to this point in time.

REFERENCES

Das, B.M., 1978. Model tests for uplift capacity of foundations in clay. Soils and Foundations 18 (2), 17–24.

Das, B.M., 1980. A procedure for estimation of ultimate uplift capacity of foundations in clay. Soils and Foundations 20 (1), 77–82.

Das, B.M., 1990. Earth Anchors. Elsevier, Amsterdam.

Khatri, V.N., Kumar, J., 2009. Vertical uplift resistance of circular plate anchors in clays under undrained condition. Comput. Geotech 36 (8), 1352–1369.

Mehryar, Z., Hu, Y., Randolph, M.F., 2002. Pullout capacity of circular plate anchor in clay-FE analysis. International Symposium on Numerical Models in Geomechanics-NUMOG VIII. Balkema, Rotterdam, pp. 507–513.

Merifield, R.S., Lyamin, A.V., Sloan, S.W., Yu, H.S., 2003. Three-dimensional lower bound solutions for stability of plate anchors in clay. J. Geotech. Geoenviron. Eng., ASCE 129 (3), 243–253.

Merifield, R.S., Sloan, S.W., Yu, H.S., 2001. Stability of plate anchors in undrained clay. Geotechnique 51 (2), 141–153.

Meyerhof, G.G., 1973. Uplift Resistance of Inclined Anchors and Piles. Int. Conf. Soil Mech. Found. Eng. 21 (1), 167–172.

Meyerhof, G.G., Adams, J.I., 1968. The ultimate uplift capacity of foundations. Can. Geotech. J 5 (4), 225–244.

Rao, K.S.S., Kumar, J., 1994. Vertical uplift capacity of horizontal anchors. J. of Geotech. Eng. Div., ASCE 120 (7), 1134–1147.

Rowe, R.K., Davis, E.H., 1982. The behaviour of anchor plates in sand. Geotechnique 32 (1), 25–41.

Saran, S., Ranjan, G., Nene, A.S., 1986. Soil anchors and constitutive laws. J. Geotech. Eng. 112 (12), 1084–1100.

Song, Z., Hu, Y., Randolph, M.F., 2008. Numerical simulation of vertical pullout of plate anchors in clay. J. Geotech. Geoenviron. Eng. 134 (6), 866–875, doi:10.1061/(ASCE)1090-0241(2008)134:6(866).

Thorne, C.P., Wang, C.X., Carter, J.P., 2004. Uplift capacity of rapidly loaded strip anchors in uniform strength clay. Geotechnique 54 (8), 507–517.

Vesic, A.S., 1971. Breakout resistance of objects embedded in ocean bottom. J. Soil Mech. Found. Div., ASCE 97 (9), 1183–1205.

Wang, D., Hu, Y., Randolph, M.F., 2010. Three-dimensional large deformation finite-element analysis of plate anchors in uniform clay. J. Geotech. Geoenviron. Eng. 136 (2), 355–365, doi:10.1061/(ASCE)GT.1943-5606.0000210.

CHAPTER 5

Vertical Anchor Plates in Cohesionless Soil

5.1 INTRODUCTION

Anchors help outwardly-directed loads exerted on the foundation of a structure to go to a greater depth. Vertical anchors are a subcategory of plate anchors, and resist the horizontal loading in construction of sheet pile walls. Fig. 5.1 illustrates a vertical anchor plate and each of its geometric parameters. These parameters consist of h, B, H which are representatives of height, width and depth of embedment of an anchor plate, respectively.

It is worth noting that, due to the flexibility of sheet piles, the lateral earth pressure of them is inconsistent with the lateral pressure estimated by the Rankine or Coulomb earth pressure theories. A lot of research has been published regarding vertical anchors. Rowe (1952) mainly concentrated on the estimation of holding capacity and displacement of plate anchors. Rowe (1952) illustrated that the movement of approximately 0.1% of an anchor contributes to the elongation of its tie rod because vertical plate anchors are connected to the wall by a tie rod. This amount of movement may lead to a reduction of the bending moment of sheet pile wall.

There are two different anchor conditions: shallow (ie, the condition in which the embedment ratio H/h is relatively small) and deep (ie, the condition of the greater embedment ratio). It is worth mentioning that the holding capacity of a vertical anchor plate relies upon the passive force that is exerted by the soil placing in front of the vertical anchor plate. While the passive failure surface of the soil at its ultimate load intersects the surface of the ground at shallow condition, the local shear failure appears in the deep condition. Fig. 5.2 illustrates the failure surface that emerged in sand that is placed in front of a shallow square plate anchor (from Hueckel's (1957) observations).

Design and Construction of Soil Anchor Plates.
DOI: http://dx.doi.org/10.1016/B978-0-12-420115-6.00005-9

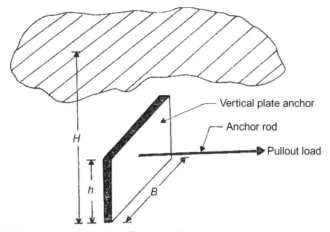

Figure 5.1 Geometric parameters of a vertical plate anchor.

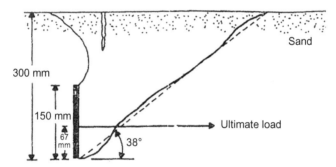

Figure 5.2 Failure surface in front of a square anchor slab embedded in sand.

The effects of contributing factors in determination of the ultimate holding capacity are:
- Embedment ratio, H/h ratio
- Width to height ratio, B/h
- Shear strength parameters of the soil, φ and c
- Friction angle at the anchor–soil interface.

This chapter investigates the function of anchors in sand.

5.2 VERTICAL ANCHORS IN COHESIONLESS SOIL

5.2.1 Rankine's Theory Regarding to Lateral Earth Pressure

Teng (1962) is one of the researchers who proposed a novel method for estimation of the ultimate holding capacity of vertical anchors by applying the

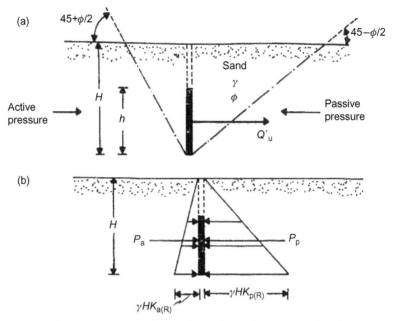

Figure 5.3 Ultimate holding capacity of strip vertical anchor as derived by Teng (1962).

lateral earth pressure theory of Rankine. Fig. 5.3a illustrates the failure surface in a granular soil around a vertical anchor locating in a shallow condition (the condition in which the h/H ratio is less than 1/3 to 1/2) at the ultimate load.

As shown in Fig. 5.3b the ultimate capacity of a vertical anchor can be calculated as below:

$$Q'_u = P_p - P_a$$

$$P_p = \frac{1}{2}\gamma H^2 K_{P(R)}$$

$$P_a = \frac{1}{2}\gamma H^2 K_{a(R)}$$

$$\tan^2\left(45 + \frac{\phi}{2}\right)$$

$$\tan^2\left(45 - \frac{\phi}{2}\right)$$

$$N = 2 \int_0^H \left(\frac{H-z}{H} \right) \left[H\sqrt{K_{P(R)}} + H\sqrt{K_{a(R)}} \right] (dz)(\gamma K_O)$$

$$= \frac{1}{3} K_O \gamma \left[\sqrt{K_{P(R)}} + \sqrt{K_{a(R)}} \right] H^3$$

$$F = N \tan \phi = \frac{1}{3} K_O \left[\sqrt{K_{P(R)}} + \sqrt{K_{a(R)}} \right] H^3 \tan \phi$$

$$Q_u = Q'_u B + F = B(P_p - P_a) + \frac{1}{3} K_O \left[\sqrt{K_{P(R)}} + \sqrt{K_{a(R)}} \right] H^3$$

$$\phi = 32°$$

$$K_{P(R)} = \tan^2 \left(45 + \frac{\phi}{2} \right) = \tan^2 \left(45 + \frac{32}{2} \right) = 3.25$$

$$K_{a(R)} = \tan^2 \left(45 - \frac{\phi}{2} \right) = \tan^2 \left(45 - \frac{32}{2} \right) = 0.307$$

$$P_p = \frac{1}{2} \gamma H^2 K_{P(R)} = \frac{1}{2} (105)(4)^2 (3.25) = 2730 \text{ lb/ft}$$

$$P_a = \frac{1}{2} \gamma H^2 K_{a(R)} = \frac{1}{2} (105)(4)^2 (0.307) = 257.9 \text{ lb/ft}$$

$$F_3 = \frac{1}{3} K_O \gamma \left[\sqrt{K_{P(R)}} + \sqrt{K_{a(R)}} \right] H^3 \tan \phi$$

$$F_3 = \frac{1}{3} (0.4)(105) \left[\sqrt{3.25} + \sqrt{0.307} \right] (4)^3 \tan 32 = 699.25 \text{ lb}$$

$$Q_u = B(P_p - P_a) + F = (5)(2730 - 257.9) + 699.25 = 13.060 \text{ lb}$$
$$P_{a(H)} = P_a \cos \phi$$
$$P_{a(V)} = P_a \sin \phi$$

$$P_{p(H)} = \frac{1}{2}\gamma H^2 K_{PH}$$

$$P_{P(V)} = \frac{1}{2}\gamma H^2 K_{PH} \tan \delta$$

$$P_{a(V)} + W = P_{p(V)}$$

$$Q'_u = R_{ov} Q'_{u(B)}$$

$$R_{ov} = \frac{C_{ov} + 1}{C_{ov+\frac{H}{h}}}$$

$$Q_u = Q'_u B_e S_f$$

$$S = \infty$$

$$F' \leq F$$

The parameters applied in the aforementioned equations are as follows:

Q'_u, The ultimate holding capacity per unit width of the vertical anchor

P_p, The passive force per unit width of the vertical anchor

P_a, The active force per unit width of the vertical anchor

γ, Unit weight of soil

ϕ, Soil friction angle

$K_{P(R)}$, $\tan^2\left(45 + \frac{\phi}{2}\right)$ which is Rankine passive earth coefficient

$K_{a(R)}$, $\tan^2\left(45 - \frac{\phi}{2}\right)$ which is Rankine active earth coefficient.

5.2.2 Teng's Theory

It is worth noting that in the case of anchors with limited width, it is crucial to consider the frictional resistance. When calculating frictional resistance, the calculation of earth pressure normal is the first priority. Teng (1962) proposed the following procedure for determination of normal earth pressure (Fig. 5.4).

$$N = 2 \int_0^H \left(\frac{H-z}{H}\right) \left[H\sqrt{K_{P(R)}} + H\sqrt{K_{a(R)}}\right](dz)(\gamma K_O)$$

$$= \frac{1}{3} K_O \gamma \left[\sqrt{K_{P(R)}} + \sqrt{K_{a(R)}}\right] H^3$$

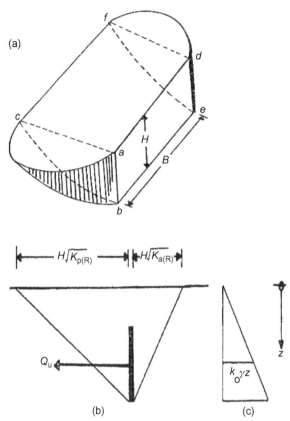

Figure 5.4 Frictional resistance developed along the vertical faces of the failure surface according to Teng's method.

$$F = N \tan \phi = \frac{1}{3} K_O \left[\sqrt{K_{P(R)}} + \sqrt{K_{a(R)}} \right] H^3 \tan \phi$$

$$Q_u = Q'_u B + F = B(P_p - P_a) + \frac{1}{3} K_O \left[\sqrt{K_{P(R)}} + \sqrt{K_{a(R)}} \right] H^3$$

In the above equations, K_O is the earth pressure coefficient at rest, F is the frictional resistance and Q_u is the ultimate holding capacity.

Example 5.1

Determine the ultimate holding capacity for a vertical anchor with the following values: $H = 2$ ft, $B = 2.5$ ft, $\gamma = 100$ lb/ft^3, $\phi = 32°$.

Solution

$$K_{P(R)} = \tan^2\left(45 + \frac{\phi}{2}\right) = \tan^2\left(45 + \frac{32}{2}\right) = 3.25$$

$$K_{a(R)} = \tan^2\left(45 - \frac{\phi}{2}\right) = \tan^2\left(45 - \frac{32}{2}\right) = 0.307$$

$$P_p = \frac{1}{2}\gamma H^2 K_{P(R)} = \frac{1}{2}(100)(2)^2(3.25) = 650 \text{ lb/ft}$$

$$P_a = \frac{1}{2}\gamma H^2 K_{a(R)} = \frac{1}{2}(100)(2)^2(0.307) = 61.4 \text{ lb/ft}$$

$$F_3 = \frac{1}{3}K_O\gamma\left[\sqrt{K_{P(R)}} + \sqrt{K_{a(R)}}\,\right]H^3 \tan\phi$$

$$F_3 = \frac{1}{3}(0.4)(100)\left[\sqrt{3.25} + \sqrt{0.307}\,\right](2)^3 \tan 32 = 83.244 \text{ lb}$$

$$Q_u = B(P_p - P_a) + F = (2.5)(650 - 61.4) + 83.244 = 1471.5 \text{ lb}$$

5.3 DEVELOPMENTS ON DETERMINATION OF VERTICAL ANCHOR CHARACTERISTICS EMBEDDED IN COHESIONLESS SOIL

Research has been conducted in an attempt to determine the ultimate holding capacity of vertical anchors. Researchers also sought a proper model to estimate the failure surface of sand around the vertical anchors. We will now briefly evaluate some of these studies.

5.3.1 Evaluation of Ovesen and Storeman's (1972) Analysis

In 1972, Ovesen and Storeman conducted an investigation into determination of holding capacity per unit width of anchors.

This study was mainly founded on three major types of plate anchors:
1. Continuous anchors, which also are known as the basic case (Fig. 5.5a)
2. Anchors that are embedded in the depth of H with the height of h (Fig. 5.5b)
3. Anchors with limited width to height ratio, B/h (Fig. 5.5c).

Let us look at the basic case in more detail.

Fig. 5.6 illustrates the estimated failure surface of sand at the ultimate load in a case of a vertical anchor embedded in sand. The figure signifies that Ovesen and Stromann (1972) estimated the failure surface in front of the anchor that comprises of a straight rupture line (BC), Prandtl radial shear zone (ACD) and Rankine passive zone (ADE).

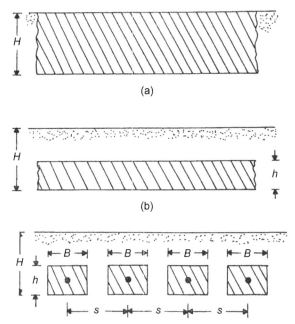

Figure 5.5 Ovesen and Stromann's (1972) analysis: (a) basic case, (b) strip case, and (c) actual case.

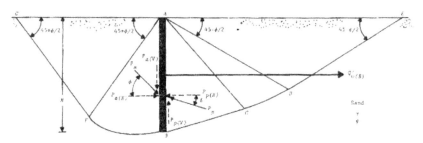

Figure 5.6 Basic case-failure surface at ultimate load.

In order to determine the ultimate holding capacity, the determination of the vertical and horizontal components of passive force in front of the anchor slab and active force is the first priority:

$$P_{a(H)} = P_a \cos \phi$$

$$P_{a(V)} = P_a \sin \phi$$

$$P_{p(H)} = \frac{1}{2}\gamma H^2 K_{PH}$$

$$P_{P(V)} = \frac{1}{2}\gamma H^2 K_{PH} \tan\delta$$

In the equations mentioned above, γ is the unit weight of soil, K_{PH} is the horizontal component of the passive earth pressure coefficient, δ is the anchor–soil friction angle, $P_{a(V)}$ and $P_{a(H)}$ are the passive and horizontal components of P_a and $P_{P(V)}$ and $P_{p(H)}$ are the vertical and horizontal components of passive force.

In order to maintain the vertical equilibrium we can write:

$$P_{a(V)} + W = P_{p(V)}$$

For maintaining the horizontal equilibrium we can write:

$$Q'_{u(B)} = P_{P(H)} - P_{a(H)} = \frac{1}{2}\gamma H^2 K_{PH} - P_{a(H)}$$

It is worth mentioning that, according to the Fig. 5.7, K_{PH} can be determined by $K_{PH} \tan \delta$ and ϕ as the known parameters. Also, the magnitude of $P_{a(V)}$ and $P_{a(H)}$ can be estimated by any ordinary active earth pressure theory.

Now let's look at the strip case.

According to the Ovesen's experimental research, the ultimate holding capacity of a strip anchor could be determined as below:

$$Q'_u = R_{ov} Q'_{u(B)}$$

where

Q'_u, ultimate holding capacity of strip anchor

$R_{ov} = \frac{C_{ov} + 1}{C_{ov} + (H/h)}$ (Fig. 5.8 illustrates the variation of R_{ov} with h/H).

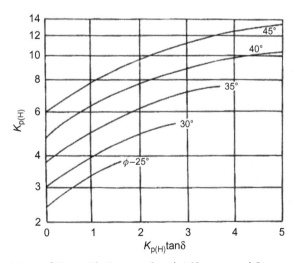

Figure 5.7 Variation of $K_{p(H)}$ with $K_{p(H)} \tan \delta$ and ϕ (Ovesen and Stromann, 1972).

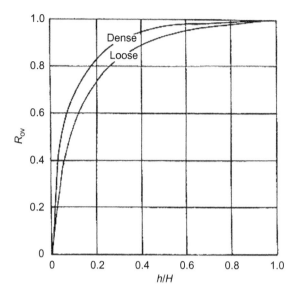

Figure 5.8 Variation of R_{ov} with H/h based on Ovesen and Stromann's theory (1972).

Note that C_{ov} is 19 for dense sand and 14 for loose sand.

And finally, the limited B/h ratio.

Ovesen and Stromann's research signified that in the case of using anchors with limited width to height ratio, the shape of the failure surface would be three-dimensional, which contributes to the below equation:

$$Q_u = Q'_u B + F$$

where

F, frictional resistance of the sides of three-dimensional failure surface.

Ovesen also introduced a new parameter in his research. S is the parameter which shows the center-to-center spacing of the anchors located in a row as shown in Fig. 5.9.

Due to differing amounts of S/B ratio, the three-dimensional failure surfaces may overlap, which contributes to the below equation for determination of the ultimate holding capacity.

$$Q_u = Q'_u B + F'$$

It is worth noting that in the aforementioned equation, F' is the effective frictional resistance ($F' \le F$).

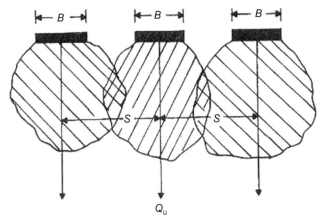

Figure 5.9 Overlapping of failure surface in soil in front of a row of vertical anchors.

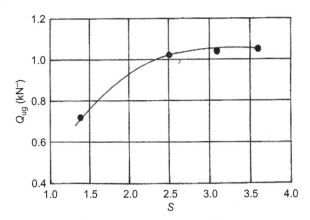

Figure 5.10 Variation of ultimate group capacity with center-to-center spacing of anchor as observed by Hueckel (1957).

5.3.2 Hueckel's Theory

Hueckel (1957) also carried out an investigation in order to determine the proper S/B ratio which contributes to equalization of F and F'. According to Fig. 5.10, for a group anchor that consists of three anchors (100 mm × 100 mm), the proper S/B ratio is approximately 3−4 which contributes to the equalization of F and F'.

5.3.3 Ovesen and Stromann's Theory

In 1972, Ovesen and Stromann presented a new model for taking the effects of overlapped failure surfaces into consideration. In this model, a new parameter, B_e as the equivalent width (as shown in Fig. 5.11) has been presented.

Fig. 5.12, which presents the variation of $(B_e - B)/(H - h)$ with $(S - B)/(H - h)$, can be applied for estimation of B_e in this model. Also, the ultimate holding capacity would be determined as follows:

$$Q_u = Q'_u B_e$$

It is worth mentioning that in the case of a single anchor $(S = \infty)$, the aforementioned equation changes to:

$$Q_u = Q'_u B_e S_f$$

where $S_f = B_e/B$, the shape for of the anchor.

According to Fig. 5.12 it can be indicated that for a single anchor $(S = \infty)$, S_f can be estimated for dense and loose sand as mentioned below:

$$S_f = 0.26 \left(\frac{1 + (H/h)}{(B/h)} \right) + 1 \quad \text{(For loose sand)}$$

$$S_f = 0.42 \left(\frac{1 + (H/h)}{(B/h)} \right) + 1 \quad \text{(For dense sand)}$$

According to the aforementioned equations, the ultimate holding capacity of a single anchor can be determined by the following equation:

$$Q_u = B \left[\frac{1}{2} \gamma H^2 K_{PH} - P_{a(H)} \right] \left(\frac{C_{ov} + 1}{C_{ov} + (H/h)} \right) \left[F \left(\frac{(H/h) + 1}{(B/h)} \right) + 1 \right]$$

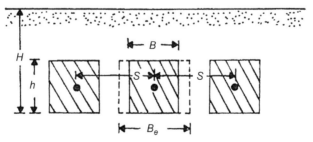

Figure 5.11 Definition of equivalent width.

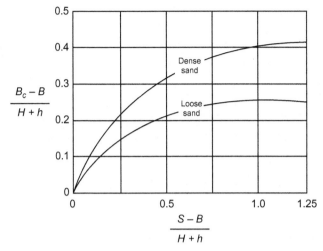

Figure 5.12 Variation of $(B_e - B)/(H + h)$ with $(S - B)/(H + h)$.

where

$$\text{For loose sand: } \begin{array}{l} C_{ov} = 19 \\ F = 0.42 \end{array}$$

$$\text{For dense sand: } \begin{array}{l} C_{ov} = 14 \\ F = 0.26 \end{array}$$

5.3.4 Evaluation of Meyerhof's Study (1965)

The study of Sokolovskii (1965) paved the way for Meyerhof's study. Meyerhof applied the presented passive and active coefficients which have been proposed in Sokolovskii's research and showed the following relationship for calculation of the ultimate holding capacity per unit width of a strip anchor. Note that in the following equation, K_b is the pullout coefficient whose variation is shown in Fig. 5.13.

$$Q_u' = \frac{1}{2}\gamma H^2 K_b$$

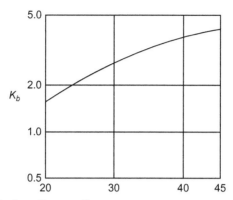

Figure 5.13 Meyerhof's pullout coefficient.

By combining the aforementioned equations, the following equation has been presented for estimation of the ultimate holding capacity:

$$4 < \frac{H}{h} < 7$$

$$Q'_u = \gamma h^2 \left\{ \left[K_{P(R)} - K_{a(R)} \right] \left(\frac{H}{h} - \frac{1}{2} \right) + \left[\frac{K_{P(R)} \sin 2\phi}{2 \tan \left(45 + (\phi/2) \right)} \right] \left(\frac{H}{h} - 1 \right)^2 \right\}$$

$$F_q = \frac{Q'_u}{\gamma h H} = \frac{h}{H} \left\{ \left[K_{P(R)} - K_{a(R)} \right] \left(\frac{h}{H} - \frac{1}{2} \right) \right.$$
$$\left. + \left[\frac{K_{P(R)} \sin 2\phi}{2 \tan \left(45 + (\phi/2) \right)} \right] \left(\frac{H}{h} - 1 \right)^2 \right\}$$

$$F_q = \frac{Q_u}{\gamma (h B) H} = F_{q(\text{strip})} + \phi \left(\frac{h}{B} \right) \left(\frac{h}{H} \right) \left[\sqrt{K_{P(R)}} - \sqrt{K_{a(R)}} \right] \left(\frac{H}{h} - \frac{2}{3} \right)$$
$$+ \frac{1}{2} (1 + \phi) \left(\frac{h}{B} \right) \left(\frac{h}{H} \right) K_{P(R)} \sin 2\phi \left(\frac{H}{h} - 1 \right)$$

While

$$\text{For loose sand:} \quad \begin{array}{l} C_{ov} = 19 \\ F = 0.42 \end{array}$$

$$\text{For dense sand:} \quad \begin{array}{l} C_{ov} = 14 \\ F = 0.26 \end{array}$$

5.3.5 Evaluation of Biarez, Boucraut, and Negre's Study (1965)

The main aim of the research of Biarez et al. (1965) was to analyze the magnitude of the ultimate holding capacity in different embedment ratios (H/h). According to this analysis, in an embedment ratio of $(H/h) \prec 4$ the variants in determination of the ultimate holding capacity are W_a (weight of the anchor) and δ (anchor soil friction angle).

Many years later, Dickin and Leung (1985) asserted that at the embedment ratio of $4 \prec (H/h) \prec 7$, the results of the investigation of Biarez et al. are more accurate. Dickin and Leung (1985) presented the below simplified equation for estimation of ultimate holding capacity of a strip anchor.

$$Q_u' = \gamma h^2 \left\{ \left[K_{P(R)} - K_{a(R)} \right] \left(\frac{H}{h} - \frac{1}{2} \right) + \left[\frac{K_{P(R)} \sin 2\phi}{2 \tan \left(45 + (\phi/2) \right)} \right] \left(\frac{H}{h} - 1 \right)^2 \right\}$$

In a nondimensional form:

$$F_q = \frac{Q_u'}{\gamma h H} = \frac{h}{H} \left\{ \left[K_{P(R)} - K_{a(R)} \right] \left(\frac{h}{H} - \frac{1}{2} \right) \right.$$
$$\left. + \left[\frac{K_{P(R)} \sin 2\phi}{2 \tan \left(45 + (\phi/2) \right)} \right] \left(\frac{H}{h} - 1 \right)^2 \right\}$$

where in the preceding equation:

F_q, breakout factor

$K_{P(R)}$, Rankine passive earth pressure coefficient

$K_{a(R)}$, Rankine active earth pressure coefficient.

Dickin et al. presented an equation for calculation of the ultimate resistance of a shallow single anchor as follows:

$$F_q = \frac{Q_u}{\gamma(hB)H} = F_{q(strip)} + \phi \left(\frac{h}{B} \right) \left(\frac{h}{H} \right) \left[\sqrt{K_{P(R)}} - \sqrt{K_{a(R)}} \right] \left(\frac{H}{h} - \frac{2}{3} \right)$$
$$+ \frac{1}{2} (1 + \phi) \left(\frac{h}{B} \right) \left(\frac{h}{H} \right) K_{P(R)} \sin 2\phi \left(\frac{H}{h} - 1 \right)$$

5.3.6 Evaluation of Neely, Stuart, and Graham's Study (1965)

The stress characteristics analysis of Sokolovskii (1965) was the foundation of the research of Neely, Stuart and Graham (1973) that contributed to

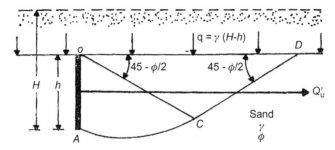

Figure 5.14 Surcharge method of analysis by Neely et al. (1973).

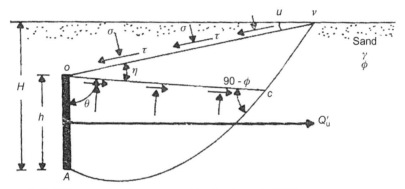

Figure 5.15 Failure mechanism assumed by Neely et al. (1973).

the presentation of a model for determination of the holding capacity of vertical strip anchor. The theoretical investigations led to presenting the force coefficient $M_{\gamma q}$ by the surcharge method and equivalent free surface methods.

1. **Surcharge method:** The name of this method is the representative of the general idea of it, which signifies that the soil placed on top of the anchor can be assumed as a simple surcharge of $q = \gamma(H - h)$. The failure surface comprises a logarithmic spiral zone and a straight line as illustrated in Fig. 5.14.

2. **Equivalent free surface method:** As shown in Fig. 5.15, the failure surface in soil around an anchor consists of a logarithmic spiral zone and a straight line which is the equivalent free surface. The research, which was aimed at predicting the ultimate bearing capacity of foundations, was conducted by Meyerhof (1951) and contributed to the presentation of free surface for the first time.

Along this equivalent free surface, the shear stress can be calculated as follows:

$$\tau = m\sigma \tan \phi$$
$$0 \prec m \prec 1$$
$$M_{\gamma q} = \frac{Q'_u}{\gamma h^2}$$
$$\frac{h}{H} = 1 + \frac{\sin \alpha \cdot \cos \phi \cdot e^{\theta \cdot \tan \phi}}{\cos (\phi + \eta)}$$
$$F_q = M_{\gamma q}\left(\frac{h}{H}\right)$$

In the aforementioned equation:

σ, Effective normal stress

ϕ, Soil friction angle

m, Mobilization factor $(0 \prec m \prec 1)$.

Note: In the presented failure mechanism, the embedment ratio and the nondimensional force coefficient $(M_{\gamma q})$ can be determined as follows:

$$\frac{h}{H} = 1 + \frac{\sin \alpha \cdot \cos \phi \cdot e^{\theta \cdot \tan \phi}}{\cos (\phi + \eta)}$$

$$M_{\gamma q} = \frac{Q'_u}{\gamma h^2}$$

The relationship between the nondimensional force coefficient and embedment ratio can be estimated by the below equation:

$$F_q = M_{\gamma q}\left(\frac{h}{H}\right)$$

Fig. 5.16 illustrates the variation of nondimensional force coefficient estimated by the surcharge method. It is worth noting that the magnitude of nondimensional force coefficient according to the surcharge method is a function of δ and ϕ (soil anchor friction angle and soil friction angle, respectively).

Also, according to Fig. 5.17, it could be indicated that the magnitude of nondimensional force coefficient based on the equivalent free surface method is a function of mobilization factor, m.

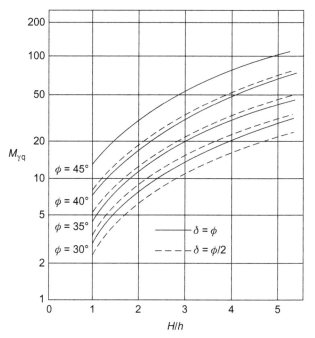

Figure 5.16 Variation of $M_{\gamma q}$ with H/h based on the surcharge method.

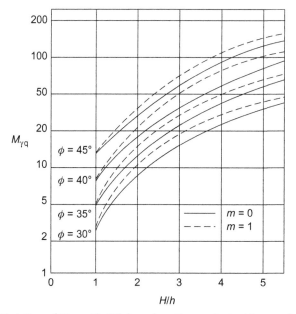

Figure 5.17 Variation of $M_{\gamma q}$ with H/h based on the equivalent free surface method.

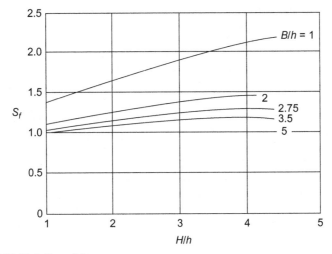

Figure 5.18 Variation of S_f.

For a single anchor, the force coefficient can be calculated by the preceding equation although a nondimensional shape factor, which was determined by Neely et al. (1973) and according to Fig. 5.18 is a function of embedment ratio (H/h) and width to height ratio (B/h), should be incorporated in the aforementioned equation.

$$M_{\gamma q} = \frac{Q'_u}{\gamma h^2 B} = M_{\gamma q(\text{strip})} S_f$$

or

$$Q_u = (\gamma B h^2) M_{\gamma q(\text{strip})} S_f$$

According to Das (1975), whose theory was founded on the Neely et al. (1973) study, the ultimate holding capacity of square anchors, which are anchors with equal width and height, could be determined as follows:

$$Q_u = C\gamma \left(\frac{H}{h}\right)^n h^3$$

It is worth mentioning that in Fig. 5.19, C is a function of friction angle, ϕ.

5.3.7 Evaluation of Kumar and Mohan Rao's Study (2002)

Kumar and Mohan Rao's (2002) study was aimed at determining the ultimate pullout capacity of vertical plate anchors. Despite a lot

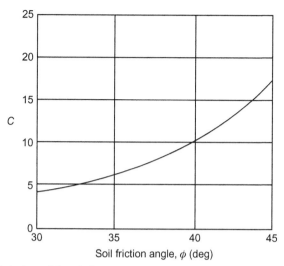

Figure 5.19 Variation of C with friction angle.

of research that had been conducted in this area, none had considered the effects of earthquake acceleration coefficient on pullout resistance. Kumar and Mohan Rao's (2002) study was founded on the concept of equivalent free surface which was presented by Meyerhof (1951) for the first time. According to this analysis, no surcharge pressure had been placed along the ground surface, thus the ultimate pullout capacity presented by Kumar (2002) was estimated by the below equation:

$$P_u = 0.5\gamma h^2 F_\gamma$$

where

h, the height of the anchor plate

F_γ, nondimensional pullout capacity factor

γ, unit weight of the soil.

The results of Kumar's analysis contributed to the presentation of diagrams which illustrate the variation of F_γ with the embedment ratio (H/h) for different given values of horizontal earthquake acceleration coefficient (α_h), δ/ϕ, and ϕ (Figs. 5.20−5.22). Note that δ and ϕ are friction angle of soil mass and the friction angle of the interface between the front surface of the anchor and the surrounding soil mass, respectively.

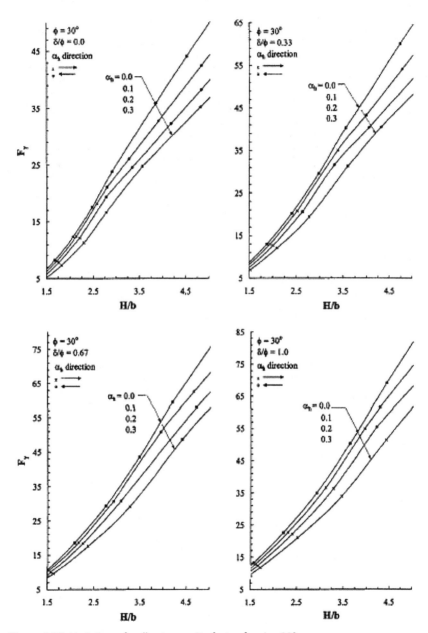

Figure 5.20 Variation of pullout capacity factor for $\phi = 30°$.

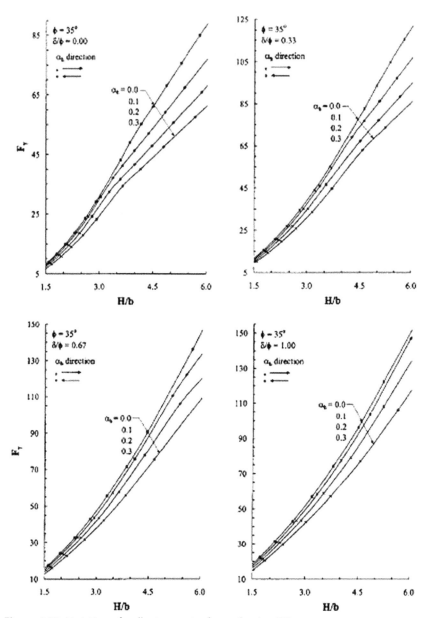

Figure 5.21 Variation of pullout capacity factor for $\phi = 35°$.

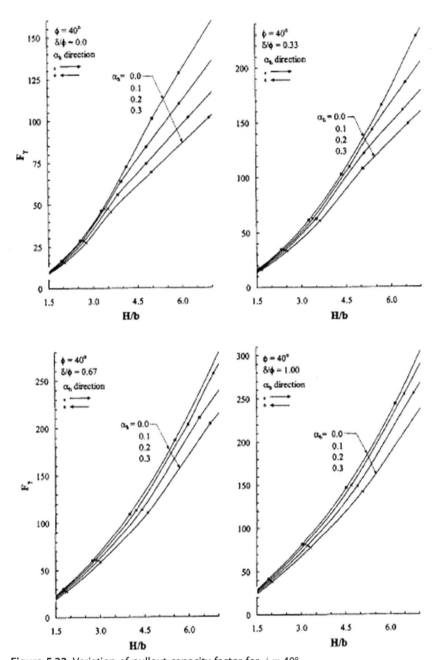

Figure 5.22 Variation of pullout capacity factor for $\phi = 40°$.

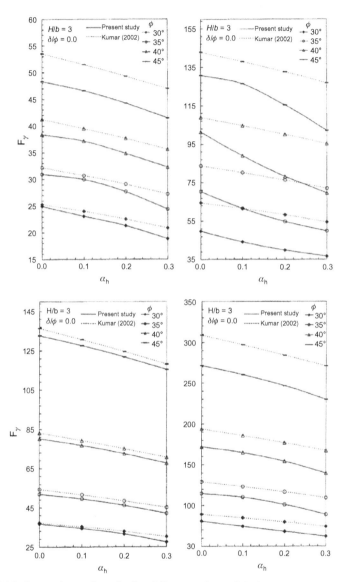

Figure 5.23 Comparison of results for different values of (α_h).

Only Kumar's (2002) study considered the impact of a pseudo–static horizontal earthquake on the horizontal holding capacity of plate anchor. Fig. 5.23 illustrates the comparison which have been made between Kumar's research and this study (Kumar and Mohan Rao's (2002)). This comparison signifies that the present pullout factor is less than the estimated pullout factor of Kumar's (2002) study.

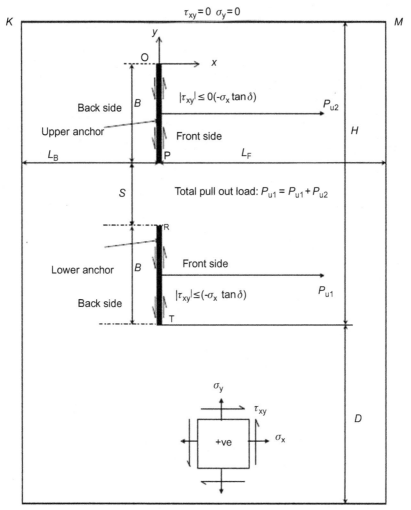

Figure 5.24 Definition of anchors' location.

5.3.8 Evaluation of Bhattacharya and Kumar's Study (2012)

Bhattacharya and Kumar (2012) have conducted a study into the ultimate holding capacity of a group of two vertical strip anchors embedded in sand (Fig. 5.24).

These two vertical anchors are located parallel to each other. In fact, Bhattacharya and Kumar intended to investigate the effect of various H/B, δ/ϕ and ϕ on the magnitude of the total group horizontal failure load (P_{ut}).

In this study, Bhattacharya asserted that the relationship between pullout capacity force factor and the pullout capacity (F_γ) for a single anchor is:

$$F_\gamma = \frac{P_u}{\gamma BH}$$

where

P_u, the magnitude of the horizontal collapse load per unit length of the strip plate

H, the depth of the lower edge of the plate below the ground surface

B, the width of each anchor plate.

According to Bhattacharya's study, the variation of F_γ with different rates of H/B for known δ/ϕ and ϕ have been estimated (Fig. 5.25).

Example 5.2

Recalculate Example 5.1 using the theory of Ovesen and Stromann. Assume $W = 0$.

Solution

Calculation of $P_{a(H)}$ and $P_{a(V)}$

$$P_{a(H)} = \frac{1}{2}\gamma H^2 K_a \cos \phi$$

$$P_{a(V)} = \frac{1}{2}\gamma H^2 K_a \sin \phi$$

For $\phi = 32°$, $K_a \approx 0.28$

$$P_{a(H)} = \frac{1}{2}(100)(2)^2(0.28) \cos 32° = 49.865 \text{ lb/ft}$$

$$P_{a(V)} = \frac{1}{2}(100)(2)^2(0.28) \sin 32° = 27.82 \text{ lb/ft}$$

Calculation of K_{PH}

$$K_{PH} \tan \delta = \frac{P_{a(V)} + W}{(1/2)\gamma H^2}$$

Assume $W \approx 0$

$$K_{PH} \tan \delta = \frac{27.82}{(0.5)(100)(2)^2} = 0.1391$$

According to Fig. 5.9, for $\phi = 32°$ and, the value of $K_{PH} \approx 3.4$
Calculation of Q_u

$$Q_u = B\left[\frac{1}{2}\gamma H^2 K_{PH} - P_{a(H)}\right]\left(\frac{C_{ov} + 1}{C_{ov} + (H/h)}\right)\left[F\left(\frac{(H/h) + 1}{(B/h)}\right) + 1\right]$$

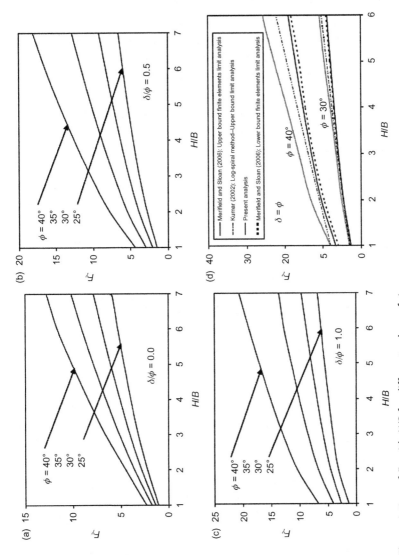

Figure 5.25 The variation of F_γ with H/B for different values of ϕ.

Assume loose sand condition; so $C_{ov} = 14$ and $F = 0.26$ thus

$$Q_u = (2.5)\left[\frac{1}{2}(100)(2)^2(3.4) - 27.82\right]\left(\frac{14+1}{14+2}\right)\left[0.26\left(\frac{2+1}{2.5}\right) + 1\right] = 2005.45 \text{ lb}$$

Example 5.3

Recalculate Example 5.2 using Meyerhof's procedure.

Solution

$$Q_u = B\left[\frac{1}{2}\gamma H^2 K_{PH} - P_{a(H)}\right]\left(\frac{C_{ov}+1}{C_{ov}+(H/h)}\right)\left[F\left(\frac{(H/h)+1}{(B/h)}\right) + 1\right]$$

From Fig. 5.15, $K_b \approx 2.95$

$$Q_u = (2.5)\left[\frac{1}{2}(100)(2)^2(2.95)\right]\left(\frac{14+1}{14+2}\right)\left[0.26\left(\frac{2+1}{2.5}\right) + 1\right] = 1904.96 \text{ lb}$$

Example 5.4 Recalculate Example 5.2 using the procedure of Neely et al. (1973) by using both the (a) equivalent free surface method and (b) the surcharge method.

Solution
Part a:

$$Q_u = \gamma Bh^2[M_{\gamma q(\text{strip})}]S_f$$

Use $m = 0$ from Fig. 5.19, for $\phi = 32°$ and $H/h = 2$, $M_{\gamma q} \approx 9.9$. Also for $B/h = 2.5$, $H/h = 2$ and $S_f \approx 1.1$ (from Fig. 5.20) so

$$Q_u = [(100)(2.5)(1)^2](9.9)(1.1) = 2722.5 \text{ lb}$$

Part b:
Assuming $\delta = \phi/2$, Fig. 5.18 gives $M_{\gamma q} \approx 7.2$ so

$$Q_u = [(100)(2.5)(1)^2](7.2)(1.1) = 1980 \text{ lb}$$

5.4 CHARACTERISTICS OF THE PASSIVE PRESSURE DISTRIBUTION OF THE SOIL AROUND A SHALLOW VERTICAL ANCHOR (HANNA, DAS, AND FORIERO, 1988)

The magnitude of the passive pressure in front of a vertical anchor plays a dominant role in the determination of the holding capacity of vertical anchors. Therefore, the nature of passive pressure distribution comes to the fore. Fig. 5.26 illustrates the distribution of the passive pressure by

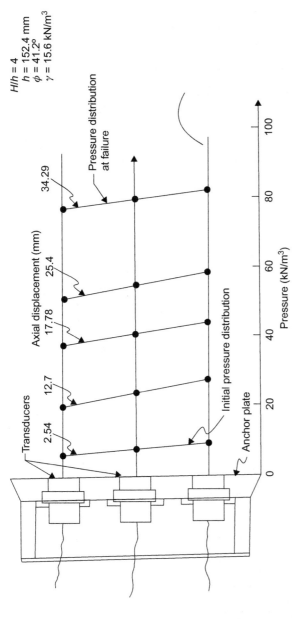

Figure 5.26 Nature of passive pressure distribution in front of a shallow vertical anchor.

means of measuring the horizontal displacement as conducted by Hanna, Das, and Foriero (1988). According to the experimental tests of Hanna et al. the parameters were as follows:

$$\gamma = 15.6 \text{ KN/m}^3 \quad \phi = 41.2°$$

$$H/h = 4 \quad h = 152.4 \text{ mm}$$

5.4.1 Hueckel, Kwasniewski, and Baran's (1965) Theory

Another group of researchers who have conducted an experimental test regarding the measurement of pressure distribution are Hueckel et al. (1965). The test was carried out on a square anchor plate embedded in sand. The results signify that the accurate distribution of passive pressure does not follow the classic pattern. According to Fig. 5.27 the parameters of the test conducted by Hueckel et al. are as follows:

$$\gamma = 16.38 \text{ KN/m}^3 \quad \phi = 34.2°$$

$$H/h = 2.5 \quad B = h = 300 \text{ mm}$$

$$\text{Horizontal displacement} = 70 \text{ mm}$$

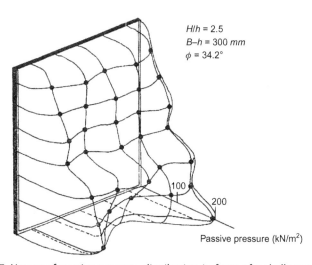

Figure 5.27 Nature of passive pressure distribution in front of a shallow anchor.

5.4.2 Hanna, Rahman, and Ayadat's (2011) Theory

Hanna et al. (2011) have conducted a study in an attempt to empirically predict the passive pressure distribution with respect to the depth of embedment and compaction of the soil in front of a vertical anchor that occurs during the vertical anchor plate's installation. According to the numerical models developed by Hanna et al. on a retaining wall (shown in Fig. 5.28), it has been concluded that the angle of shearing resistance and overconsolidation ratio of sand discernibly affect the passive earth pressure in front of the plate anchor.

Hanna et al. (2011) applied the following equation for the estimation of the passive earth pressure coefficient (K_p):

$$K_p = \frac{P_{ult}}{(1/2)h \cos \delta (2\gamma_s H'' + \gamma_s h)}$$

where

P_{ult}, ultimate load

h, height of the anchor plate

δ, friction angle between soil and the anchor plate

H'', depth of top edge of the anchor plate from the ground surface

γ_s, soil unit weight.

According to Hanna et al.'s study, it can be indicated that as the friction angle (φ) and consolidation of surrounding soil (OCR) increases, the magnitude of K_p increases considerably (Fig. 5.29). The reason of this effect is the fact that greater residual lateral pressure at greater depth contributes to the overconsolidation of sand.

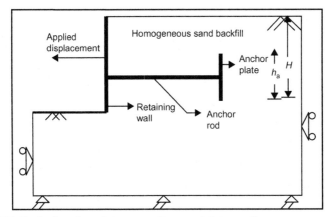

Figure 5.28 Vertical anchor plate supporting a retaining wall.

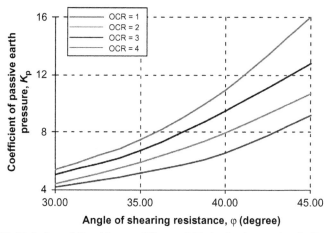

Figure 5.29 Variation of K_p with u at different OCR (embedment depth $H = 3.0$ m).

Note that the magnitude of passive pressure can be estimated by the aforementioned equation of Rankine method. Also, Hanna et al. presented the failure mechanism as illustrated in Fig. 5.30. It is worth noting that for a given embedment depth and height of the anchor, the eccentricity (e_y) and rotational angle (λ) could be determined as follows:

$$e_y = \left[\frac{H''h + (1/3)h^2}{2H'' + h}\right]$$

and

$$\lambda = \cos^{-1}\left(\frac{K_p}{K_q}\right)$$

where

    H'', Depth of top edge of the anchor plate from the ground surface
    h, height of the plate anchor
    K_q, coefficient of earth pressure due to surcharge
    λ, rotational angle.

5.4.3 Deep Vertical Anchors

As described before, the breakout factor (F_q) increases as the embedment ratio increases (Fig. 5.31). The magnitude of F_q increases and an embedment ratio, known as critical embedment ratio $(H/h)_{cr}$, remains constant. The maximum magnitude of F_q at critical embedment ratio is

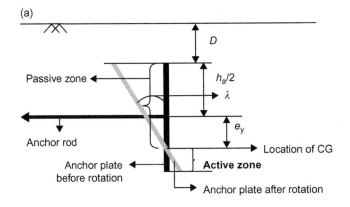

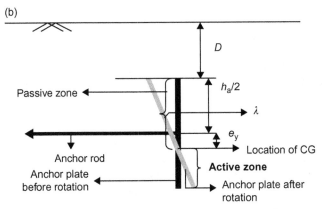

Figure 5.30 Variation of passive zone around plate anchor with increasing depth of embedment. (a) Small depth of embedment, D and (b) large depth of embedment, D.

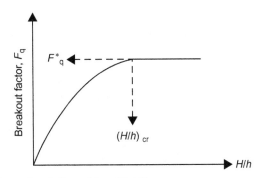

Figure 5.31 Nature of variation of F_q with H/h.

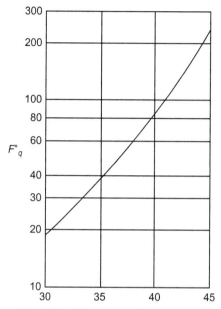

Figure 5.32 Variation of F_q^* with soil friction angle.

denoted by F_q^*. The anchors which are located at a depth less than critical depth $(H/h < (H/h)_{cr})$ are known as deep anchors.

According to Fig. 5.32 which illustrates the Ovesen's (1964) study, the following equation for rectangular anchors has been presented:

$$F_q^* = \frac{Q_u}{\gamma(Bh)H} = S_f K_o e^{\pi \tan \phi} \cdot \tan^2\left(45 + \frac{\phi}{2}\right) \cdot d_c$$

where

$$K_o = 1 - \sin \phi$$

$$d_c = 1.6 + 4.1 \tan^4 \phi$$

$$S_f = 1 + 0.2 \frac{h}{B}$$

However, Ovesen (1964) asserted that for all deep anchors, the shape factor should be considered 1. Fig. 5.32 illustrates the diagram of F_q^* considering the shape factor 1.

According to Meyerhof's (1973) study, which looked at deep anchors as well as shallow anchors, the ratio of $Q_u/\gamma(Bh)(H - h/2)$ has been asserted to be a function of friction angle as Fig. 5.33 shows. Meyerhof showed that

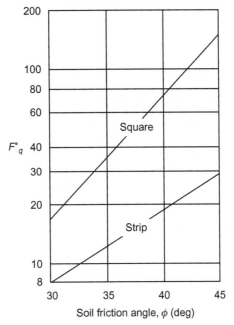

Figure 5.33 Meyerhof's value of F_q^*.

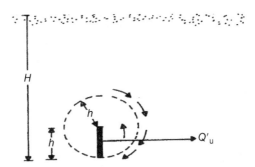

Figure 5.34 Failure mechanism around deep anchors.

the ratio of $Q_u/\gamma(Bh)(H - h/2)$ to $Q_u/\gamma(Bh)H$ is about 1.11 while the embedment ratio (H/h) is 5. So, the following equation could be used:

$$F_q^* = \frac{Q_u}{\gamma(Bh)(H - h/2)} = \frac{Q_u}{\gamma(Bh)H}$$

Biarez et al. (1965) conducted a study into the characteristics of deep strip anchors $(H/h > 7)$ in a rotational mechanism as illustrated in Fig. 5.34.

The following solution has been given for estimation of breakout factor for strip anchors:

$$F^*_{q(rectangular)} = 4\pi \left(\frac{h}{H}\right)\left(\frac{H}{h} - 1\right)\tan a \left(1 + 0.2\frac{h}{B}\right)$$

Q_u

Δ_u

$$\frac{Q}{Q_u} = \frac{(\Delta/\Delta_u)}{0.15 + 0.85(\Delta/\Delta_u)} = \frac{0.278}{0.15 + (0.85)(0.278)} = 0.72$$

$\dfrac{\Delta}{\Delta_u}$

$1 \leq \Delta/\Delta_u = 1/3.6 \leq 5$

$Q = (0.72)(Q_u) = (0.72)(16,880) = 12,154 \, lb$

ϕ_{peak}

$$S_f = 0.42\left(\frac{(H/h) + 1}{(B/h)}\right) + 1$$

$$K_p = \frac{P_{ult}}{(1/2)h \cos \delta \, (2\gamma_s H'' + \gamma_s h)}$$

(φ)

According to the Biarez et al.(1965) analysis, for calculation of the breakout factor for rectangular anchors, a shape factor parallel with what have been mentioned in Ovesen's theory should be incorporated in the preceding equation:

$$F^*_{q(rectangular)} = 4\pi \left(\frac{h}{H}\right)\left(\frac{H}{h} - 1\right)\tan \phi \left(1 + 0.2\frac{h}{B}\right)$$

In comparison between the breakout factor estimated by Ovesen and Meyerhof's studies, it is indicated that the breakout factor obtained from Ovesen's solution is higher than Meyerhof's solution. Besides, according to Fig. 5.35, the breakout factor estimated by Meyerhof's equation is higher for square anchors ($B = h$) than strip anchors.

5.5 LOAD–DISPLACEMENT RELATIONSHIP

5.5.1 Neely et al. Theory (1973)

Some design restrictions result in limited horizontal displacements. Neely et al. (1973) conducted an experimental study whose results signifies that

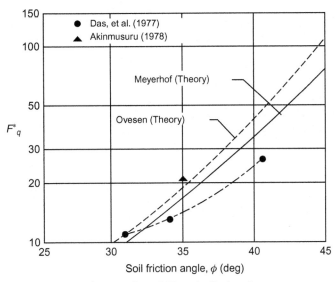

Figure 5.35 Comparison of Ovesen's and Meyerhof's theories.

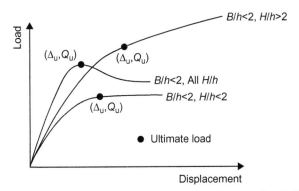

Figure 5.36 Typical nature of load versus displacement diagram for shallow anchors.

three types of load–displacement diagrams, as shown in Fig. 5.36, exist for vertical anchors located in sand:

- $B/h < 2$ and $H/h < 2$: The load increases as the horizontal displacement increases. When it reaches to maximum level, it remains constant.
- $B/h < 2$ and $H/h > 2$: The load increases as the horizontal displacement increases. When it reaches to maximum level, it becomes linear.
- $B/h > 2$ at all values of H/h: The load increases as the horizontal displacement increases. When it reaches to maximum level, it decreases.

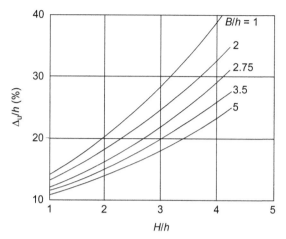

Figure 5.37 Nondimensional plot of Δ_u/h versus H/h for various values of B/h.

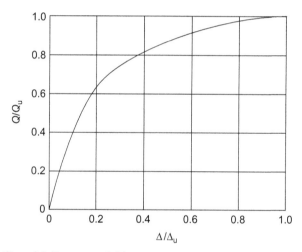

Figure 5.38 Plot of Q/Q_u versus Δ/Δ_u.

It is worth noting that the maximum load denoted by Q_u in diagrams, and its corresponding displacement has been shown by Δ_u and the magnitude of Δ_u which have been obtained from the experimental tests by Neely et al. (1973) have been shown in Fig. 5.37.

5.5.2 Das and Seeley Theory (1975)

According to Fig. 5.38 which shows the relationship between Q/Q_u and Δ/Δ_u, some discrepancies have been observed between the figurative model and the theoretical model which have been presented by Das and Seeley (1975). This model indicates that for $1 \leq B/h \leq 5$ and $1 \leq H/h \leq 5$ the load–displacement relationship can be estimated by the following equation:

$$\frac{Q}{Q_u} = \frac{(\Delta/\Delta_u)}{0.15 + 0.85(\Delta/\Delta_u)}$$

where
Δ = displacement at load Q.

5.5.3 Hanna, Rahman, and Ayadat (2011)

Hanna et al.'s research, founded on Chin's method, estimated the relationship between load and displacement as below:

$$P = \frac{\Delta}{\overline{A}\Delta + \overline{B}}$$

where
Δ, Displacement of an anchor
$\overline{A}$, slope of the straight line $(\overline{A} = 1/P_{ult})$
$\overline{B}$, Material stiffness
P, load at the displacement Δ.

Hanna et al. also conducted a comparison between data obtained from the preceding formula and empirical data. The results signify that theoretical data conform to empirical data relatively well (Fig. 5.39).

Example 5.5
Refer to Example 5.2.
1. Determine the anchor displacement at ultimate load, Q_u.
2. Determine the allowable load Q for an anchor displacement of 1 inch.

Solution
Part a: Referring to Fig. 5.31, for $B/h = 2.5$ and $H/h = 2$, $\Delta/\Delta_u \approx 0.15$ so

$$\Delta_u = 0.15(2 \times 12) = 3.6 \text{ in}$$

Part b: $\Delta = 1$ in. $\Delta/\Delta_u = 1/3.6 = 0.278$

$$\frac{Q}{Q_u} = \frac{(\Delta/\Delta_u)}{0.15 + 0.85(\Delta/\Delta_u)} = \frac{0.278}{0.15 + (0.85)(0.278)} = 0.72$$

$$Q = (0.72)(Q_u) = (0.72)(16,880) = 12,154 \text{ lb}$$

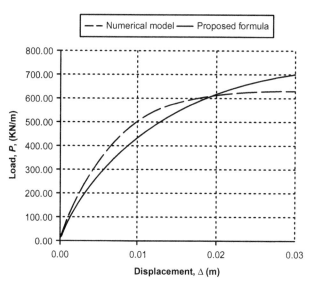

Figure 5.39 Comparison of load—displacement curves ($\varphi = 35°$ and OCR = 3).

5.6 FRICTION ANGLE AS A DESIGN CONSIDERATION

The contributing effects of the soil friction angle on determining the ultimate holding capacity of an anchor, requires accurate consideration of ϕ. Discrepancies in the results obtained from model tests and prototypes means that the scale factor of the model plays a contributing role on the achieved results. In fact, as the magnitude of the height of an anchor increases, the magnitude of the ultimate holding capacity decreases significantly. On most applications, the friction angle is not the maximum friction angle ϕ_{peak}. According to Fig. 5.40, it could be considered that, as the stress level increases, the friction angle becomes less, thus the peak value of friction angle occurs in at lower stress level. In fact, a higher peak value could be estimated for the model rather than the prototype.

5.6.1 Shape Factor as a Design Consideration

Ovesen and Stromann (1972) presented a model for predicting the shape factor for a single anchor in a case where the soil is compacted after the placement of the anchor in soil. The recommended equation is:

$$S_f = 0.42 \left(\frac{(H/h) + 1}{(B/h)} \right) + 1$$

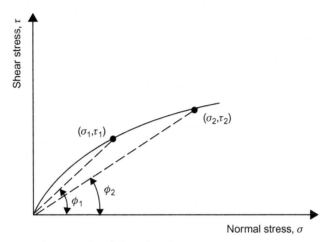

Figure 5.40 Curvilinear Mohr's failure diagram.

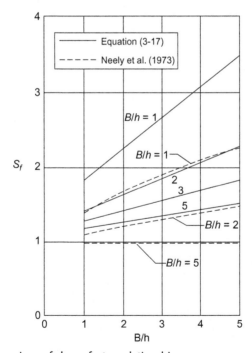

Figure 5.41 Comparison of shape factor relationship.

In comparison with the recommended solution by Neely et al. (1973), Ovesen and Stromann's suggested shape factor is relatively higher than Neely's estimated shape factor (Fig. 5.41). This is assumed to be the result of simplification that Neely considered in his estimation of shape factor. In fact, in Neely's model it was assumed that the behavior of $B/h = 5$ is the same as an strip anchor.

5.7 GENERAL RECOMMENDATIONS

As described in preceding sections, many different factors may affect the ultimate holding capacity of anchors. However, based on the empirical experiences, some recommendations can be made.

Due to the fact that conducting the plane strain test in not quite relevant for determination of friction angle, it is suggested that the tri-axial peak friction angle be calculated. However, the results of the friction angles determined by plane stress test and tri-axial test do not conform to each other totally. In fact, $\phi_{peak(triaxial)}$ is ten percent higher than $\phi_{peak(plane-strain)}$.

In the next step, the peak friction angle can be applied for determination of the holding capacity of a single anchor by Ovesen and Stromann's method as follows:

$$Q'_u = \frac{1}{2}\gamma H^2 (K_{PH} - K_a \cos\phi)\left(\frac{C_{ov} + 1}{C_{ov} + (H/h)}\right)$$

$$F_{q(rectangular)} = \frac{Q'_u}{\gamma h H} = 0.5\left(\frac{H}{h}\right)(K_{PH} - K_a \cos\phi)\left(\frac{C_{ov} + 1}{C_{ov} + (H/h)}\right)$$

$$\times \left[F\left(\frac{(H/h) + 1}{(B/h)}\right) + 1\right] \le F^*_{(rectangular)-Meyerhof}$$

$$\delta = \frac{\phi}{2}$$

$$Q_{all(field)} = \frac{0.3\ Q_u}{2} = 0.15\ Q_u$$

$$\frac{Q}{Q_u} = 1/2\frac{(\Delta/\Delta_u) = 0.13}{0.15 + 0.85(\Delta/\Delta_u)}$$

$$\Delta \le 0.03\ h$$

$$Q'_u = \frac{1}{2}\gamma H^2 (K_{PH} - K_a \cos\phi)\left(\frac{C_{ov} + 1}{C_{ov} + (H/h)}\right)$$

or

$$F_q = \frac{Q_u'}{\gamma h H} = 0.5\left(\frac{H}{h}\right)(K_{PH} - K_a \cos \phi)\left(\frac{C_{ov} + 1}{C_{ov} + (H/h)}\right)$$

The maximum value of F_q should be equal to $F^*_{q(strip)}$ presented by Meyerhof (1973):

$$0.5\left(\frac{H}{h}\right)(K_{PH} - K_a \cos \phi)\left(\frac{C_{ov} + 1}{C_{ov} + (H/h)}\right) \leq F^*_{q(strip)-Meyerhof}$$

Also the preceding equations for calculation of F_q should be as follows:

$$F_{q(square)} = \frac{Q_u'}{\gamma h H} = 0.5\left(\frac{H}{h}\right)(K_{PH} - K_a \cos \phi)\left(\frac{C_{ov} + 1}{C_{ov} + (H/h)}\right)$$

$$\times \left[F\left(\frac{H}{h} + 1\right) + 1\right] \leq F^*_{(square)-Meyerhof}$$

Also for rectangular anchors:

$$F_{q(rectangular)} = \frac{Q_u'}{\gamma h H} = 0.5\left(\frac{H}{h}\right)(K_{PH} - K_a \cos \phi)\left(\frac{C_{ov} + 1}{C_{ov} + (H/h)}\right)$$

$$\times \left[F\left(\frac{(H/h) + 1}{(B/h)}\right) + 1\right] \leq F^*_{(rectangular)-Meyerhof}$$

It is worth noting that at $H/h < 3$, the holding capacity of anchors can also be determined by surcharge method with $\delta = \phi/2$ or the equivalent free surface method with $m = 0$ as presented by Neely et al. (1973):

$$Q_u = \gamma h^2 B M_{\gamma q(strip)} \cdot S_f$$

Neely et al. also asserted that the ultimate holding capacity of a single anchor estimated in the preceding section should be decreased by 31% for 10-fold increase in size due to the scale effects. Also, Dickin and Leung (1985) recommended that a decrease of 30% in Q_u along with tri-axial peak friction angle might be suitable:

$$Q_{u(field)} = 0.3 \, Q_u$$

It is also possible to consider a safety factor of 2 for allowable load:

$$Q_{all(field)} = \frac{0.3 \, Q_u}{2} = 0.15 \, Q_u$$

According to the aforementioned equations:

$$\frac{Q}{Q_u} = \frac{\Delta/\Delta_u}{0.15 + 0.85(\Delta/\Delta_u)}$$

When the ratio of $Q/Q_u = 1/2$
Then

$$\frac{\Delta}{\Delta_u} = 0.13$$

which means that a deflection of $\Delta = 0.13 \, \Delta_u$ will contribute to a value of $\Delta < 0.065$ h for square anchors and $\Delta \leq 0.03$ h for strip anchors.

REFERENCES

Bhattacharya, P., Kumar, J., 2012. Horizontal pullout capacity of a group of two vertical strip anchors plates embedded in sand. Geotech. Geol. Eng. 30, 513–521.

Biarez, I., Boucraut, L.-M., Negre, R., 1965. Limiting Equilibrium of Vertical Barriers Subjected to Translation and Rotation Forces. Proceedings of the 6th International Conference on Soil Mechanics and Foundation Engineering, Vol. II. Montreal, Canada, Sept, pp. 368–372.

Chin F.K., 1972. The inverse slope as a prediction of ultimate bearing capacity of piles. In: Proceedings of the 3rd southeast Asian conference on soil engineering, Hong Kong, pp. 83–91.

Das, B.M., 1990. Earth Anchors. Elsevier, Amsterdam.

Das, B.M., Seeley, G.R., 1975. Load displacement relationship for vertical anchor plates. J. Geotech. Eng. Div. ASCE. 101 (7), 711–715.

Dickin, E., Leung, C.F., 1985. Evaluation of design methods for vertical anchor plates. Geotech. Eng. ASCE 111 (4), 500–520.

Hanna, A., Rahman, F., Ayadat, T., 2011. Passive earth pressure on embedded vertical plate anchors in sand. Acta Geotech. 6, 21–29.

Hanna, A.M., Das, B.M., Foriero, A., 1988. Behavior of shallow inclined plate anchors in sand. Special Topics in Foundations. Geotech. Spec. Tech. Pub. ASCE. 16, 54–72.

Hueckel, S. (1957) Model tests on anchoring capacity of vertical and inclined plates. In: Proc. IV Intl. Conf. Soil Mech. Found. Eng, London, England, vol. 2, pp. 203–206.

Hueckel, S. & Kwasniewski, J. & Baran, L. (1965). Distribution of passive earth pressure on the surface of a square vertical plate embedded in soil. In: Proc. IV Intl. Conf. Soil Mech. Found. Eng. London, England, vol. 2, pp. 203–206.

Kumar, J., Mohan Rao, V.B.K., 2002. Seismic bearing capacity factors for spread foundations. Geotechnique 52 (2), 79–88.

Kumar, J., 2002. Seismic horizontal pullout capacity of vertical anchors in sands. Canadian Geotechnical Journal 39 (4), 982–991.

Meyerhof, G.G., 1951. The ultimate bearing capacity of foundations. Geotechnique. 2 (4), 301–332.

Meyerhof, G.G. (1973) Uplift resistance of inclined anchors and piles. In: Proc. VIII Intl. Conf. Soil Mech. Found. Eng, Moscow, USSR, vol. 2(1), pp. 167–172.

Neely, W.J., Stuart, J.G., Graham, J., 1973. Failure loads of vertical anchor plates in sand. J. Soil Mech. Found. Div. ASCE. 99 (9), 669–685.

Ovesen, N.K. (1964). Anchor Slabs, Calculation Methods and Model Tests. Copenhagen, Denmark.

Ovesen, N.K., Stromann, H., 1972. Design Method for Vertical Anchor Slabs in Sand. Proceedings of Specialty Conference on Performance of Earth and Earth-Supported Structures, Vol. 1—2, pp. 1418—1500.

Rankine, W., 1857. On the stability of loose earth. Philos. Trans. R. Soc. Lond. 147.

Rowe, P.W. (1952) Anchored sheet pile walls. In: Proc. Institute of Civil Engineers, London, England, vol. 1(1), pp. 27—70.

Sokolovskii, V.V., 1965. Statics of Granular Media. Pergamon Press, New York.

Teng, W.C., 1962. Foundation Design. Prentice-Hall, Englewood Cliffs, NJ.

CHAPTER 6

Vertical Anchor Plates in Cohesive Soil

6.1 INTRODUCTION

As we read in chapter "Vertical Anchor Plates in Cohesionless Soil," anchors are used to transfer an outward load to a greater depth and withstand any tensile forces. In fact, anchors' capacity of withstanding load is the passive resistance of soil. Anchors are commonly used to provide the required uplift resistance and the required lateral resistance.

Vertical anchors are an important type of anchor that can be used in either sand or clay soils. Vertical anchor plates are mostly utilized to bear the pressure in seawalls, sheet piles, and retaining walls. Many diverse factors affect the ultimate capacity which an anchor can resist. These factors comprise of H/h (embedment ratio), B/D (width to height ratio) and friction angle of soil−anchor interface. The most often used shapes of anchor plates for vertical systems are square $(B=D)$, circular, rectangular $(B \neq D)$, and strip $(B/D \geq 6)$. Rowe (1952) described the importance of soil anchor movement when the anchor is less than 0.1% of the height sheet wall as it can decrease the bonding moment. In this chapter we will look at vertical plates that are located in clay (cohesive) soils.

The plan in this chapter is to assess the effects of contributing factors in determining the ultimate holding capacity of vertical anchor(s) located in clay. Dickin and Leung (1983) conducted a study which revealed that the anchor's geometric factors play a contributing role in its capacity and displacement. These factors are:
- Embedment ratio, H/h ratio
- Width to height ratio, B/h
- Shear strength parameters of the soil, φ and c
- Friction angle at the anchor−soil interface.

It is worth noting that many researchers proposed various theories on vertical anchor plates such as Terzaghi (1943), Hansen (1953), Ovesen (1964), Ovesen and Stromann (1972), Biarez et al. (1965), Meyerhof (1973),

Design and Construction of Soil Anchor Plates.
DOI: http://dx.doi.org/10.1016/B978-0-12-420115-6.00006-0

Neely et al. (1973), Das and Seely (1975), Rowe and Davis (1982), Hanna et al. (1988), Murray and Geddes (1989), Basudhar and Singh (1994), Merifield et al. (2006), Niroumand et al. (2010), and Bhattacharya and Kumar (2012). Also, in order to determine the ultimate holding capacity of an anchor embedded in clay, Ranjan and Arora (1980) and Das et al. (1975) have conducted many laboratory tests. Most of this research has been conducted on the determination of the ultimate holding capacity of a single anchor, although some researchers such as Sahoo and Kumar (2012) have carried out a study into investigation of the ultimate holding capacity of a group of two vertical anchors. Published research varies: some are numerical analyses, others are laboratory based. As the application of earth anchors expands, there is a need for more investigation into different types of anchors.

6.2 ULTIMATE HOLDING CAPACITY

Fig. 6.1 illustrates a regular vertical anchor located in soil. The general parameters of such anchors embedded in saturated clay are h, B and H which are representative of height, width of the anchor and depth of embedment, respectively. Also, the undrained shear strength of clay is denoted by c_u. The prominent attribute of clay is the value of its friction angle, which is zero.

6.2.1 Mackenzie's Theory

The ultimate holding capacity of an anchor plate, ultimate holding capacity per unit weight, and the ultimate holding capacity for an

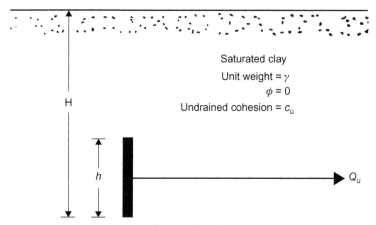

Figure 6.1 Geometric parameters of vertical anchor embedded in saturated clay.

anchor embedded in clay in a nondimensional form (breakout factor), can be expressed as follows:

$$Q'_u = \frac{Q}{B}$$

$$F_c = \frac{Q'_u}{c_u h}$$

where

Q'_u = ultimate holding capacity per unit width
Q_u = ultimate holding capacity of anchor plate
F_c = breakout factor

Although some research has been carried out regarding the ultimate holding capacity of vertical anchor plates located in undrained clay, there are some empirical tests which have been carried out on strip anchors (plane strain condition).

These laboratory tests have been conducted by Mackenzie (1995) on strip anchors located in two different clay soils. According to the empirical results of Mackenzie's study, the relationship between the breakout factor and the embedment ratio (H/h) can be plotted as seen in Fig. 6.2. The figure shows that the breakout factor increases as the embedment ratio of the vertical anchor also increases. The maximum limit of the breakout factor is denoted by F_c^* which occurs at critical embedment ratio $(H/h)_{cr}$, and thereafter the magnitude of the breakout remains constant.

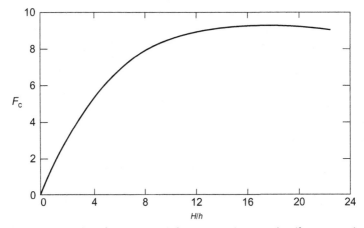

Figure 6.2 Average plot of F_c versus H/h for strip anchors in clay (friction angle = 0).

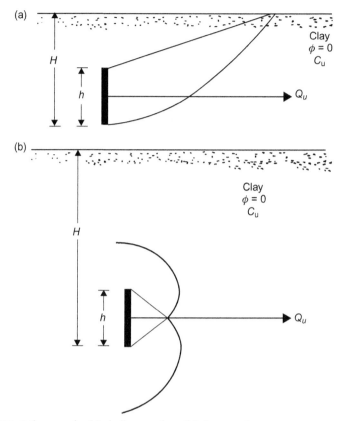

Figure 6.3 Failure mode: (a) shallow anchor; (b) deep anchor.

This indicates that the failure mode in sand is divided into two different regions. The first region occurs in the case of embedment ratios that are less than critical embedment ratio $(H/h \leq (H/h)_{cr})$. Anchors located in this region are called shallow anchors and surrounding soil failure take place at ultimate load (Fig. 6.3). In the laboratory results of Mackenzie's tests, the magnitude of F_c^* was estimated at 9. Also, Mackenzie calculated the approximate value of $(H/h)_{cr}$ as 12.

6.2.2 Meyerhof's Theory

Meyerhof (1973) also conducted a series of laboratory tests that contributed to presentation of the below estimated values of the critical embedment ratio and the breakout factor. Meyerhof presented these estimates for square and strip anchors. Meyerhof's estimates for square anchors are as follows:

$$F_c = 1.2\left(\frac{H}{h}\right) \leq 9$$

$$\left(\frac{H}{h}\right)_{cr} = 7.5$$

And for strip anchors:

$$F_c = 1.0\left(\frac{H}{h}\right) \leq 8$$

$$\left(\frac{H}{h}\right)_{cr} = 8$$

Although Mackenzie's study paved the way for further research, not all questions were answered. For instance, researchers were skeptical about the reliance of $(H/h)_{cr}$ and B/h on the undrained shear strength of the plate or the parameters which affect F_c of rectangular anchors. In an attempt to answer this, Das et al. (1985) carried out many laboratory studies in order to clarify the vagueness regarding to the aforementioned parameters. Das et al.'s research was aimed at investigating the dependence of breakout factor F_c of square anchors on undrained shear strength of the clay (Fig. 6.4). Fig. 6.4 illustrates that critical embedment ratio of clay increases as c_u increases. However, the value of c_u remains constant after a maximum value. The below relationship can be presented for estimation of the mentioned behavior:

$$\left(\frac{H}{h}\right)_{cr-s} = 4.7 + 2.9 \times 19^{-3} c_u \leq 7$$

where:

$\left(\frac{H}{h}\right)_{cr-s}$ = critical embedment ratio of square anchors (which means $h = B$) and c_u is lb/ft^2.

c_u = undrained shear strength of clay (kN/m^2)

It is noting that according to Meyerhof's study the maximum of $\left(\frac{H}{h}\right)_{cr-s}$ was estimated at 7.

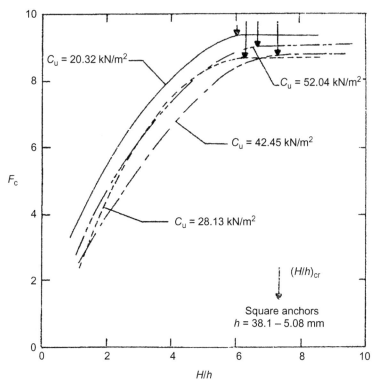

Figure 6.4 Model test results for Das et al. the variation of F_c with H/h for square anchors.

Das et al. estimated a relationship between the critical embedment ratio of square and rectangular anchors. This relationship can be expressed as follows:

$$\left(\frac{H}{h}\right)_{cr-s} = 4.7 + 0.0606\,c_u \leq 7 \quad \text{(For square anchors)}$$

$$\frac{\left(\dfrac{H}{h}\right)_{cr-R}}{\left(\dfrac{H}{h}\right)_{cr-s}} = \left[0.9 + 0.1\left(\frac{B}{h}\right)\right] \leq 1.31 \quad \text{(For rectangular anchors)}$$

According to Das et al.'s study, the following equation has been presented for estimation of breakout factor of deep rectangular anchors. Note that in the following equation, $F^*_{c(R)}$ and $F^*_{c(S)}$ are used to express the breakout factor of deep rectangular anchors and breakout factor of deep square anchors, respectively.

$$F^*_{c(R)} = F^*_{c(S)}\left[0.825 + 0.175\left(\frac{h}{H}\right)\right]$$

Also, for shallow anchors, the following equation can be applied:

$$F_c = \frac{F^*_c\left(\dfrac{H}{h}\right)}{0.41\left(\dfrac{H}{h}\right)_{cr} + 0.59\left(\dfrac{H}{h}\right)}$$

6.2.3 Shahoo and Kumar's Theory

Many researchers have conducted numerical and laboratory studies in order to determine the ultimate holding capacity of single vertical anchors embedded in clay, but only a few research groups have investigated determining the ultimate holding capacity of a group of vertical anchors comprises of two vertical anchors located in clay. Shahoo et al. investigated the effects of a spacing factor between two vertical anchors. It has been revealed that in a maximum spacing between anchors S_{cr}, the magnitude of group failure load becomes maximum. Also, Shahoo and Kumar (2012) revealed that an increase in $\gamma H/c_o$ contributes to an increase in the magnitude of group failure load.

According to Shahoo and Kumar's investigations, it has been revealed that the critical spacing between two vertical anchors is approximately $0.7-1.2$ times the height of the anchor plate. In Fig. 6.5, each of the two anchor plates have the height of B and the bottom edge of the lower vertical anchor is located at a depth H from ground surface. Also, Shahoo and Kumar presented the following equation for estimation of the undrained shear strength due to the fact that it varies linearly with the depth (h):

$$c = c_o\left(1 + m\frac{h}{B}\right)$$

where:

c = value of cohesion at a depth of h
c_o = value of cohesion at ground surface
m = nondimensional factor

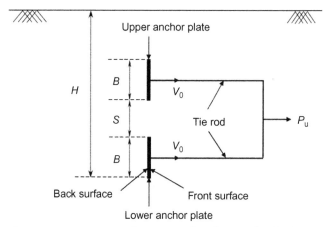

Figure 6.5 Geometric parameters of two vertical anchors embedded in clay.

Shahoo and Kumar assumed that the group of anchors are subjected to horizontal pullout load and the failure occurs in the both of the vertical plates at the same time. Shahoo and Kumar's (2012) study was aimed at determination of the total horizontal group failure load per unit weight P_u. It has been revealed that according to the below equation, the pullout factor can be determined by the failure load per unit length of the anchor and the height of the anchor (B) and the shear strength of the soil:

$$F_{c\gamma} = \frac{P_u}{Bc}$$

Fig. 6.6 illustrates the variation of $F_{c\gamma}$ with $\gamma H/c_o$ for different H/B ratios. Shahoo and Kumar's research revealed that the pullout resistance of the group anchor comprising two vertical anchors is almost $1.21-2.25$ times the resistance of a single anchor with the same H/B.

6.2.4 Merifield, Sloan and YU's Theory

Merifield et al.'s (2001) study was mostly analytical. According to Merifield et al.'s study, the undrained shear strength is estimated to vary linearly as the embedment depth of the anchor increases:

$$c_u(z) = c_{uo} + \rho z$$

where:

c_{uo} = undrained shear strength at the ground surface
z = depth below ground surface
$\rho = dc_u/dz$

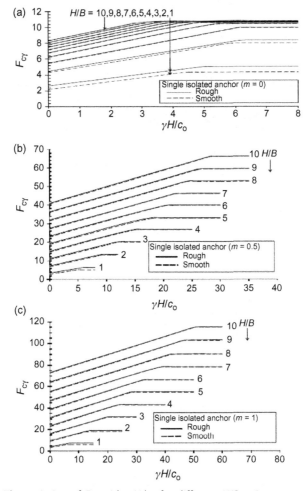

Figure 6.6 The variation of $F_{c\gamma}$ with $\gamma H/c_o$ for different H/B ratios.

According to Merifield et al.'s study, the ultimate capacity of the anchor embedded in clay (q_u) can be determined by the following equation:

$$q_u = c_u N_c$$

Note that the breakout factor is defined for homogeneous and nonhomogeneous soils according to the following equations:

For homogeneous soils:

$$N_c = \left(\frac{q_u}{c_u}\right)_{\gamma \neq 0, \rho = 0} = N_{co} + \frac{\gamma H_a}{c_u}$$

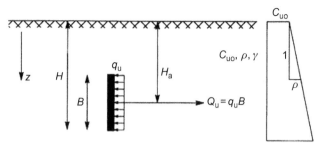

Figure 6.7 Vertical plate anchor.

And N_{co} can be defined as:

$$N_{co} = \left(\frac{q_u}{c_u}\right)_{\gamma=0, \rho=0}$$

For nonhomogeneous soils:

$$N_{co} = \left(\frac{q_u}{c_{uo}}\right)_{\gamma \neq 0, \rho \neq 0} = N_{co\rho} + \frac{\gamma H_a}{c_{uo}}$$

And $N_{co\rho}$ is defined as:

$$N_{co\rho} = \left(\frac{q_u}{c_{uo}}\right)_{\gamma=0, \rho \neq 0}$$

According to the Fig. 6.7 $H_a = H - B/2$.

Merifield et al. investigated the effects of anchor roughness and overburden pressure and increasing strength with depth. However, Merifield's research was purely analytical rather than based upon laboratory tests.

6.3 THE PROCEDURE FOR ESTIMATION OF THE ULTIMATE LOAD

Based on the results that have been gathered by several research groups (Mackenzie, Meyerhof, Shahoo and Kumar, Merifield, Sloan and YU, and related researchers) at the present time, the below procedure has been developed for estimating the ultimate holding capacity of single rectangular anchors that are located in clay. Remember, clay is a type of soil with no friction angle.

1. In the first step, the proper embedment ratio (H/h) and the width to height ratio (B/h) should be estimated.
2. The value of shear strength should be calculated (C_u).

3. In the next step according to the below equation, the critical embedment ratio for the square anchor should be determined:

$$\left(\frac{H}{h}\right)_{cr-s} = 4.7 + 2.9 \times 19^{-3} \, c_u \leq 7$$

4. According to the estimation regarding to B/h made in the first step, the ratio of $\dfrac{\left(\dfrac{H}{h}\right)_{cr-R}}{\left(\dfrac{H}{h}\right)_{cr-s}}$ should be calculated.

5. Based on steps 3 and 4, the value of $\left(\dfrac{H}{h}\right)_{cr-R}$ can be estimated.

6. In the next step, it should be determined whether the anchor is deep or shallow. Thus, the embed ratio estimated in the first step should be compared with $\left(\dfrac{H}{h}\right)_{cr}$. Two general categories exist in this step:

a. $\dfrac{H}{h} > \left(\dfrac{H}{h}\right)_{cr}$: then the anchor is a deep anchor:

$$F^*_{c(R)} = 9\left[0.825 + 0.175\left(\frac{h}{B}\right)\right]$$

Or

$$Q_u = 9 c_u Bh\left[0.825 + 0.175\left(\frac{h}{B}\right)\right]$$

b. $\dfrac{H}{h} < \left(\dfrac{H}{h}\right)_{cr}$: the embedment ratio is less than critical embedment ratio which means that the anchor is a shallow anchor and the following equations are applied for determination of Q_u:

$$F^*_{c(R)} = 9\left[0.825 + 0.175\left(\frac{h}{B}\right)\right]$$

$$Q_u = c_u Bh\left[\frac{F^*_{c(R)}\left(\dfrac{H}{h}\right)}{0.41\left(\dfrac{H}{h}\right)_{cr-R} + 0.59\left(\dfrac{H}{h}\right)}\right]$$

Example 6.1

Estimate the ultimate breakout load of a rectangular anchor plate with the following details: $H = 0.8$ m, $h = 0.2$ m, $B = 0.4$ m, and c_u kN/m^2.

Solution

$$\frac{H}{h} = \frac{0.8 \text{ m}}{0.2 \text{ m}} = 4$$

$$\frac{B}{h} = \frac{0.4 \text{ m}}{0.2 \text{ m}} = 2$$

$$Q_u = c_u Bh \left[\frac{F^*_{c(R)}\left(\frac{H}{h}\right)}{0.41\left(\frac{H}{h}\right)_{cr-R} + 0.59\left(\frac{H}{h}\right)} \right]$$

$$F^*_{c(R)} = 9\left[0.825 + 0.175\left(\frac{h}{B}\right)\right] = 9\left[0.825 + 0.175\left(\frac{0.2}{0.4}\right)\right] = 8.21$$

$$Q_u = [(48)(0.4)(0.2)]\left[\frac{(8.21)(4)}{0.41(7.7) + 0.59(4)}\right] = 51.43 \text{ kN}$$

According to the following equation:

$$\left(\frac{H}{h}\right)_{cr-S} = 4.7 + 0.0606\,c_u = 4.7 + 0.0606(48) = 7.609 > 7$$

So use $\left(\frac{H}{b}\right)_{cr-s} = 7$

From the following equation we calculate $\dfrac{\left(\frac{H}{h}\right)_{cr-R}}{\left(\frac{H}{h}\right)_{cr-s}}$:

$$\frac{\left(\frac{H}{h}\right)_{cr-R}}{\left(\frac{H}{h}\right)_{cr-s}} = \left[0.9 + 0.1\left(\frac{B}{h}\right)\right] = [0.9 + 0.1(2)] = 1.1$$

And based on the previous step we calculate $\left(\frac{H}{h}\right)_{cr-s}$

So $\left(\frac{H}{h}\right)_{cr-R} = (1.1)\,(7) = 7.7$

The actual H/h is 2 so $H/h \prec \left(\dfrac{H}{h}\right)_{cr}$ and the anchor is a shallow anchor.

$$Q_u = c_u Bh \left[\frac{F_{c(R)}^* \left(\dfrac{H}{h}\right)}{0.41 \left(\dfrac{H}{h}\right)_{cr-R} + 0.59 \left(\dfrac{H}{h}\right)} \right]$$

$$F_{c(R)}^* = 9 \left[0.825 + 0.175 \left(\frac{h}{B}\right) \right] = 9 \left[0.825 + 0.175 \left(\frac{0.2}{0.4}\right) \right] = 8.21$$

So

$$Q_u = [(48)(0.4)(0.2)] \left[\frac{(8.21)(4)}{0.41(7.7) + 0.59(4)} \right] = 51.43 \text{ kN}$$

6.4 GENERAL REMARKS

Researchers are skeptical about some certain limitations and estimations applied in the results of the studies aimed at determinating the ultimate holding capacity of plate anchors embedded in clay.

Most of the estimation and equations presented in this chapter are based on laboratory test carried out on prototypes, which signifies that the scale factor has not yet been taken into consideration although it is predicted that such effects are ignorable in clay soils.

Although the most research considers only single anchors, there is a high probability that anchors are applied in groups. According to Fig. 6.8,

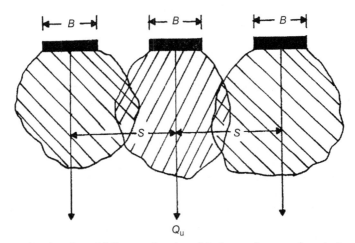

Figure 6.8 Overlapping of failure surface in soil in front of a row of vertical anchors.

the failure surface in soil around each anchor in the anchor group may contradict, contributing to a reduction in the ultimate capacity of the anchor Q_u. The reduction factor according to below equation is noted by

$$\eta: \text{Efficiency factor} = f\left(\frac{S}{B}, \frac{H}{h}\right)$$

$$Q_{u(\text{actual})} = \eta Q_{u(\text{isolated})}$$

However, the efficiency factor has not yet been determined.

A safety factor need to be considered to estimate the allowable holding capacity. This safety factor is considered to be 3.

REFERENCES

Basudhar, P.K., Singh, D.N., 1994. A generalized procedure for predicting optimal lower bound breakout factors of strip anchors. Geotechnique 44 (2), 307–318.

Bhattacharya, P., Kumar, J., 2012. Horizontal pullout capacity of a group of two vertical strip anchors plates embedded in sand. Geotech. Geol. Eng. 30, 513–521.

Biarez, I.,Boucraut, L.M., Negre, R. 1965. Limiting equilibrium of vertical barriers subjected to translation and rotation forces. Proc. VI Intl. Conf. Soil mech. Found. Engrg. Montreal, Canada. 2. pp. 368–372.

Das, B.M., 1990. Earth Anchors. Elsevier, Amsterdam.

Das, B.M., Seely, G.R., 1975. Load displacement relationship for vertical anchor plates. J. Geotech. Eng. Div. ASCE 101 (7), 711–715.

Das, B.M., Tarquin, A.J., Moreno, R., 1985. Model tests for pollout resistance of vertical anchors in clay, Civ. Eng. Pract. Design Eng., 4 (2). Pergamon Press, New York, pp. 191–209.

Dickin, E.A., Leung, C.F., 1983. Centrifugal model tests on vertical anchor plates. J. Geotech. Eng. Div. 109, 1503–1525.

Hanna, A.M., Das, B.M., Forieri, A., 1988. Behavior of shallow inclined plate anchors in sand. Special Topics in Foundations. Geotech. Spec. Tech. Pub. ASCE 16, 54–72.

Mackenzie, T.R. 1995. Strength of Deadman Anchors in Clay. M.S. Thesis. Prinston University, USA.

Meyerhof, G.G., 1973. Uplift resistance of inclined anchors and piles, Proc. VIII Intl. Conf. Soil. Mech. Found. Eng., 2 (1). USSR, Moscow, pp. 167–172.

Mierfield, R.S., Sloan, S.W., Yu, H.S., 2001. Stability of plate anchors in un-drained clay. Geôotechnique 51 (2), 141–153.

Murray, E.J., Geddes, J.D., 1989. Resistance of passive inclined anchors in cohesionless medium. Geotechnique 39 (3), 417–431.

Neely, W.J., Stuart, J.G., Graham, J., 1973. Failure loads of vertical anchor plates in sand. J. Soil Mech. Found. Div. ASCE. pp. 669–685.

Niroumand, H., Anuar Kassim, K., Nazir, 2010. Analytical and Numerical Studies of Vertical Anchor Plates in Cohesion-less Soils. EJGE.

Ovesen, N.K. 1964. Anchor Slabs, Calculation Methods and Model Tests. Copenhagen, Denmark.

Ovesen, N.K., Stromann, H., 1972. Design method for vertical anchor slabs in sand, Proc. Specialty Conf. on Performance of Earth and Earth-Supported Structures, 2 (1). ASCE, pp. 1481–1500.

Ranjan, G.G., Arora, V.B. 1980. Model studies on anchors under horizontal pull in clays. In: Proceedings of the 3rd Australia New Zealand conference on geo-mechanics. 1. Wellington. pp. 65–70.

Rowe, P.W., 1952. Anchored sheet pile walls. Proc. Inst. Civ. Eng. London, England 1 (1), 27–70.

Rowe, R.K., Davis, H., 1982. The behavior of anchor plates in sand. Geotechnique 32 (1), 25–41.

Shahoo, J.P., Kumar, J., 2012. Horizontal pullout resistance for a group of two vertical plate anchors in clays. Geotech. Geol. Eng. 30, 1279–1287.

CHAPTER 7

Inclined Anchor Plates in Cohesionless Soil

7.1 INTRODUCTION

As mentioned in previous chapters, anchors are structural elements that are applied to bear the load imposed on the foundation of a structure. Engineers mostly suggest the utilization of anchors in order to provide passive support to retaining walls, sheet piles, and bulkheads. Anchor plates are a prominent type of anchor that is used. As shown in Fig. 7.1, anchor plates are divided into three subcategories: horizontal plate anchors, vertical plate anchors, and inclined plate anchors. In previous chapters, the first two categories have been discussed. In this chapter, the characteristics of inclined anchor plates will be investigated. Inclined anchors are applied in a case when there is a need for construction of a foundation where plate anchors should be located at an inclination to the horizontal plane.

The dominant assumption regarding horizontal and inclined anchors is the fact that pullout force will be transmitted axially to anchors. Much research has been carried out into determining the ultimate pullout capacity of vertical and horizontal plate anchors. Additionally, many researchers have investigated estimating the ultimate holding capacity of inclined anchors subjected to axial pull. This chapter is aimed at reviewing and discussing the ultimate holding capacity of inclined anchors placed in sand. The primary estimation for the gross ultimate holding capacity of an inclined anchor (as seen Fig. 7.2) is as expressed below:

$$Q_{u(g)} = Q_u + W_a \cos \psi$$

where:

$Q_{u(g)}$ = gross ultimate holding capacity
Q_u = net ultimate holding capacity
W_a = self-weight of anchor

Design and Construction of Soil Anchor Plates.
DOI: http://dx.doi.org/10.1016/B978-0-12-420115-6.00007-2
153

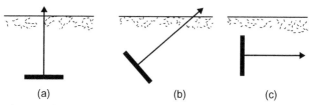

Figure 7.1 Plate anchors: (a) horizontal plate anchor; (b) inclined plate anchor; (c) vertical plate anchor.

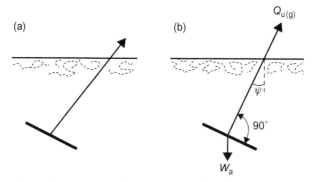

Figure 7.2 Inclined plate anchor subjected to (a) inclined pull; (b) axial pull.

7.2 EARLY THEORIES FOR INCLINED ANCHOR PLATES

7.2.1 Analysis of Harvey and Burley (1973)

The main aim of Harvey and Burley's study was determination of the ultimate holding capacity of inclined anchors, in particular shallow-inclined circular anchor plates in sand. According to the Harvey and Burley study, it was observed that shallow anchors contributed to displacement of sand at the free surface, although failure of deep anchors was not accompanied by displacement of soil surrounding the anchor plate. Fig. 7.3 shows a circular anchor plate attached to a steel rod whose diameter is denoted by h. The average depth of embedment is equal to H' and the inclination angle is equal to ψ. As illustrated in Fig. 7.3, BD and AC are the failure surfaces which are in fact arcs of two circles. These arcs intersect the surface of the ground at angles of $45 - \phi/2$ which stems from the assumptions made in Rankine's method. Also, BD and AC intersect the edges of the anchor plate at angle of $90°$ Rankine (1857).

It is worth to be noted that Balla (1961) has also presented a similar method when investigating horizontal anchor plates. Balla (1961) studied several models in dense soil and according to Balla's analysis, for shallow

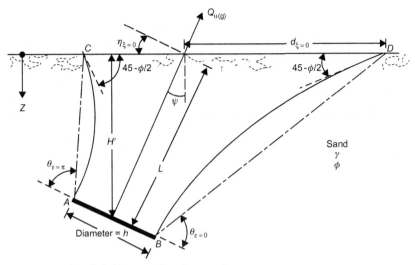

Figure 7.3 Inclined shallow circular plate anchor.

circular anchors, the failure surface could be estimated as an arc of a circle. According to Balla's study, the angle of failure surface with the horizontal is assumed to be $(45-\phi/2)$. The net ultimate uplift capacity, P_{un} for a circular anchor can be determined by the following equation:

$$P_{un} = H^3\gamma\left[F_1\left(\phi,\frac{H}{D}\right)+F_3\left(\phi,\frac{H}{D}\right)\right]$$

The failure surfaces were further simplified by two straight lines of AC and BD.

According to the results obtained from Harvey and Burley's study, the ultimate holding capacity of the plate anchor increases as the pullout capacity of shallow anchors increases.

The following equations apply to the failure zone:

$$\tan\omega = \tan\psi\cdot\sin\xi$$

$$\tan\eta = \sin\psi\cdot\cos\xi$$

$$\tan\alpha = \tan\left(45-\frac{\phi}{2}\right)\cdot\sqrt{(1+\tan^2\psi\cdot\sin^2\xi)}$$

$$\theta = \frac{\pi}{4}+\frac{\alpha}{2}+\frac{\eta}{2}$$

where:

ξ=horizontal angle locating a typical sector of the failure zone measured from the major axis of surface failure ellipse

ω=inclination angle of typical sector of failure zone relative to the failure zone axis

η=inclination angle of pullout axis to the plane perpendicular to the failure zone axis

α=angle between the curvilinear surface of sliding and horizontal surface in the plane of failure sector

The volume of the part of the failure zone having a horizontal angle of $d\xi$ is as follows:

$$d = \frac{L + \frac{h}{2} \cdot \tan \theta}{\cos \eta \cdot \tan \theta - \sin \eta}$$

$$\delta V = \left[\frac{d^2}{6} \left(L + \frac{h}{2} \cdot \tan \theta \right) \cos \eta - \frac{h^3}{48} \tan \theta \cdot \sin \eta \right] \delta \xi$$

Also, the area of a simplified curved surface of a part of the failure zone and the area of a part of the anchor plate is as follows:

$$\delta A = \frac{1}{2} \left(d + \frac{h}{2} \cdot \sec \eta \right) \left(\frac{d \cdot \cos \eta - \frac{h}{2}}{\cos \theta} \right) \delta \xi$$

$$\delta A_a = \frac{1}{8} h^2 \cdot \delta \xi$$

The lateral earth pressure at depth z is:

$$P_z = K_o \gamma z$$

where:

K_o=Coefficient of at-rest earth pressure

γ= unit weight of soil

Also, the circumferential force on the part of the failure zone caused by pressure force is:

$$\left[\frac{K_o \gamma d}{\left(L + \frac{h}{2} \cdot \tan \theta \right)} \right] \times \int_{z=0}^{z=L \cos \psi} \left(z \cdot \frac{h}{2} \cdot \tan \theta + zL - \frac{z^2}{\cos \psi} \right) dz$$

The radial force on the section can be calculated by the following equation:

$$\left[\frac{K_o \gamma dL^2 \cos^2 \psi}{2\left(L+\dfrac{h}{2}\cdot \tan \theta\right)}\right]\left[\left(\frac{h}{2}\right)\tan \theta + \frac{L}{3}\right]d\xi$$

Considering that the reaction of the anchor plate on the surrounding soil is Q at an angle of ϕ then the following equations are applicable:

$$Q \sin \theta = \gamma(\delta V)\cos \omega \cdot \sin (\phi+\theta-\eta) - F_R \cdot \cos(\theta+\phi-\eta)$$

$$Q_u = \sum_{\varsigma=0}^{\varsigma=2\pi} Q \cos \phi$$

The preceding equations can be solved by computer programs. It is worth noting that Harvey and Burley made a comparison between their analysis and the analysis conducted by Kanayan (1996) and Baker and Konder (1996). Baker and Konder also conducted several laboratory based P_u tests in order to determine the ultimate uplift capacity, P_u as mentioned in the following equations.

For shallow circular anchors

$$P_u = C_1 HD^3 r + C_2 H^3 \gamma$$

For deep circular anchors

$$P_u = 170\, D^3 \gamma + C_3 D^2 tr + C_4 HD + \gamma$$

where:
r = radius of anchor plate
t = the thickness of anchor plate
H = the depth of embedment.
C1, C2, C3, C4 are constants which are functions of angle of soil internal friction and relative density of compaction.

It is noteworthy to mention that the preceding procedure for estimation of the ultimate holding capacity is not in current use in the industry.

7.2.2 Meyerhof's Theory

Meyerhof conducted a study to determine the net ultimate holding capacity of an inclined shallow anchor. Fig. 7.4 illustrates a shallow strip anchor whose height is denoted by h and its bottom edge located at a

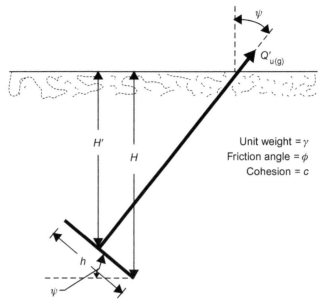

Figure 7.4 Inclined shallow strip anchor.

depth of H from the ground surface, and the average embedment depth of it is H'.

This shallow anchor is embedded in a soil, having the friction angle of ϕ and cohesion of c. The inclination angle of the aforementioned anchor with respect to horizontal is denoted by ψ. The inclined anchor is also exposed to an axial pullout force. The results of Meyerhof's study signify that for a parallel experimental condition, the pullout capacity of an inclined plate becomes more than the capacity of a vertical anchor exposed to an axial load. Meyerhof (1973) also investigated the movement of inclined anchors and a vertical anchor for a given pullout load. Results indicated that the displacement of an inclined anchor was smaller than a vertical anchor.

According to Meyerhof's study, the net ultimate holding capacity per unit width, Q'_u, can be determined by the following equation:

$$Q'_u = P_p - P_a = cK_cH + \frac{1}{2}K_b\gamma H^2$$

where:

C = cohesion

γ = unit weight of soil

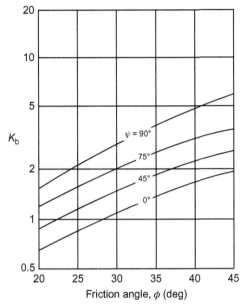

Figure 7.5 Variation of earth pressure coefficient K_b.

K_c, K_c = net earth pressure coefficients
W = weight of soil located directed above the anchor = $\gamma h H \cos \psi$

In case of substitution of weight of the soil, the preceding equation can be written as below:

$$Q'_u = cK_cH + \frac{1}{2}K_c\gamma H^2 + \gamma hH \cos^2 \psi$$

For the soil with $c=0$:

$$Q'_u = \frac{1}{2}K_b\gamma H^2 + \gamma hH \cos^2 \psi$$

Note that the value of K_b, which is the net earth pressure coefficient, can be obtained from the earth pressure coefficient estimated by Caquot and Kerisel (1949) and Sokolovskii (1965). According to Fig. 7.5, the variation of K_b with change of friction angle can be determined in a given inclination angle.

Note that the value of K_b for different friction angles can be determined by Fig. 7.6 which has also been discussed in previous chapters.

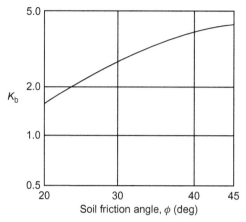

Figure 7.6 Meyerhof's pullout coefficient.

The preceding equation can be rewritten as below:

$$Q'_u = \frac{1}{2} K_b \gamma \left(H' + \frac{h \sin \psi}{2} \right)^2 + \gamma h \left(H' + \frac{h \sin \psi}{2} \right) \cos^2 \psi$$

$$F'_q = \frac{Q'_u}{\gamma h \left(H - \frac{h}{2} \right)} = \frac{Q'_u}{\gamma h H \left(1 - \frac{h}{2H} \right)} = \frac{F_q}{1 - \frac{h}{2H}}$$

where:

F'_q = average breakout factor

In the case of $\psi = 0$, the anchor is a horizontal anchor so:

$$F'_q = F_q$$

Also, with $\psi = 90°$, $H' = H$ the anchor is a vertical anchor, so:

$$F'_q = \frac{Q'_q}{\gamma h \left(H - \frac{h}{2} \right)} = \frac{Q'_u}{\gamma h H \left(1 - \frac{h}{2H} \right)} = \frac{F_q}{1 - \frac{h}{2H}}$$

where:

F_q = breakout factor

For $H/h \geq 5$, the ratio of average breakout factor and breakout factor is as follows:

$$\frac{F'_q}{F_q} \leq 1.1$$

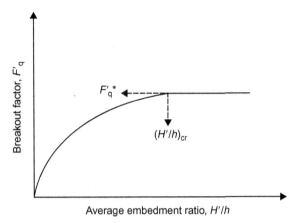

Figure 7.7 Nature of variation of F'_q with H'/h.

The passive parameters for a given soil friction angle for determination of the average breakout factor are K_b, H'/h, and the inclination angle.

As we have seen in previous chapters, a critical embedment ratio can be estimated for vertical and horizontal anchors. Thus, anchors with embedment ratios higher than the critical embedment ratio can be categorized as deep anchors. According to Fig. 7.7 the local shear failure occurs for deep anchor plates. According to diagrams presented in previous chapters, the average breakout factor remains constant beyond the critical embedment ratio, $H'/h \geq (H'/h)_{cr}$.

Meyerhof and Adams (1968) have conducted many laboratory tests on horizontal anchors in order to determine the critical embedment ratio of square anchors in loose sand and dense sand. According to the results gathered from Meyerhof and Adams's study, the critical embedment ratio for loose sand is almost 4 and for dense sand is almost 8. Also, for strip anchors the following equation can be presented:

$$\left(\frac{H'}{h}\right)_{cr-strip} \approx 1.5\left(\frac{H'}{h}\right)_{cr-square}$$

According to Fig. 7.8, which illustrates the variation of F'^*_q for a given inclination angle, $\psi = 0°, 45°, 90°$, with friction angle, the magnitude of $(H'/h)_{cr}$ increases as the inclination angle increases. Also, Meyerhof and Adams asserted that the shape factor increases with H'/h up to a critical depth.

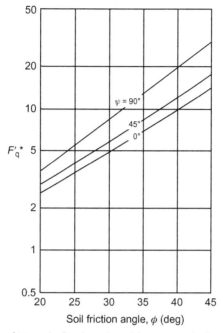

Figure 7.8 Variation of Meyerhof's $F_q^{\prime*}$ with soil friction angle for strip anchor.

The following equation can be written in order to determine the maximum breakout factor for strip anchors:

$$F_{q(\text{square})}^{\prime*} = \frac{Q_u}{\gamma A H'} = F_{q(\text{strip})}^{\prime*} \cdot S_f$$

where:

A = area of the anchor plate

S_f = shape factor

Brinch and Hansen (1961) have conducted a study that was aimed at establishing the shape factors of square vertical anchors with inclination angle of 90° which are horizontally loaded.

Fig. 7.9 illustrates the variation of $F_q^{\prime*}$ with the soil's friction angle for given inclination angle of $\psi = 0°, 90°$. Note that by interpolation between the magnitude of $F_q^{\prime*}$ for strip and square anchor, the value of $F_q^{\prime*}$ for rectangular anchors can be determined.

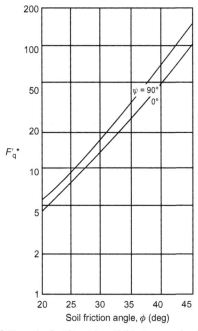

Figure 7.9 Variation of Meyerhof's $F_q'^*$ with soil friction angle ϕ for square anchor.

Example 7.1

A strip anchor is shown in Fig. 7.10. Given: $\phi = 35°$, $\gamma = 18$ kN/m^3, $h = 0.2$ m, determine the variation of the net ultimate load Q_u' for $\psi = 0°, 45°, 75°$ and $90°$.

Solution

Using the equation:

$$Q_u' = \frac{1}{2}K_b\gamma\left(H' + \frac{h\sin\psi}{2}\right)^2 + \gamma h\left(H' + \frac{h\sin\psi}{2}\right)\cos^2\psi$$

According to Fig. 7.5, the variation of K_b with the anchor inclination ψ can be estimated:

ϕ (deg)	Anchor inclination, ψ (deg)	K_b
35	0	≈ 1.4
35	45	≈ 1.8
35	75	≈ 2.7
35	90	≈ 3.9

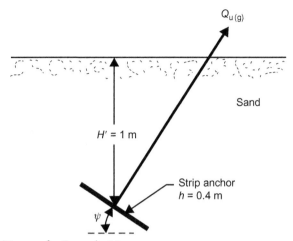

Figure 7.10 Diagram for Example 7.1.

According to the preceding equation and the values of K_b, the magnitude of Q'_u can be estimated:

$\psi = 0$

$$Q'_u = \frac{1}{2}(1.4)(18)\left(0.6 + \frac{(0.2)\sin 0}{2}\right)^2 + (18)(0.2)\left(0.6 + \frac{(0.2)\sin 0}{2}\right)\cos^2 0$$

$$= 4.536 + 2.16 = 6.696$$

$\psi = 45$

$$Q'_u = \frac{1}{2}(1.8)(18)\left(0.6 + \frac{(0.2)\sin 45}{2}\right)^2 + (18)(0.2)\left(0.6 + \frac{(0.2)\sin 45}{2}\right)\cos^2 45$$

$$= 7.162 + 1.384 = 8.546$$

$\psi = 75$

$$Q'_u = \frac{1}{2}(2.7)(18)\left(0.6 + \frac{(0.2)\sin 75}{2}\right)^2 + (18)(0.2)\left(0.6 + \frac{(0.2)\sin 75}{2}\right)\cos^2 75$$

$$= 11.649 + 0.365 = 12.014$$

$\psi = 90$

$$Q'_u = \frac{1}{2}(3.9)(18)\left(0.6 + \frac{(0.2)\sin 90}{2}\right)^2 + (18)(0.2)\left(0.6 + \frac{(0.2)\sin 90}{2}\right)\cos^2 90$$

$$= 17.199 + 0 = 17.199$$

7.2.3 Hanna et al.'s Theory

Hanna et al. (1988) have conducted an analysis of the behavior of shallow inclined plate anchors embedded sand. According Hanna et al.'s investigation, which was an analytical study, the variation of the ultimate holding capacity of shallow inclined strip anchors with different values of inclination angle between 0° and 60° have been estimated. Results signify that the pullout capacity of the anchor plate increases as the inclination angle increases. Fig. 7.11 illustrates a shallow strip anchor whose real failure surface is virtually parallel to ab' and cd'. As Fig. 7.11 illustrates,

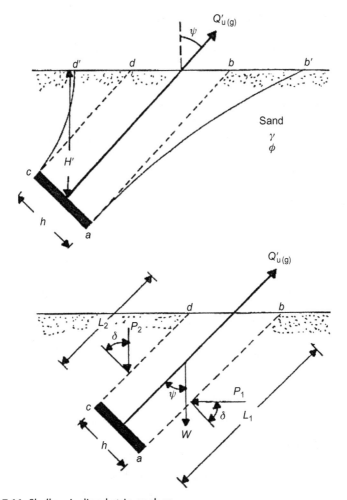

Figure 7.11 Shallow inclined strip anchor.

the passive forces per unit width of the anchor will be denoted by $P1$ and $P2$, respectively, along the planes of ab and cd. The angle between the normal of each plane and direction of each force is noted by δ. According to Fig. 7.11, the following equations can be estimated:

$$W = \frac{1}{2}(L_1 + L_2)h\gamma \cos \psi$$

$$P_1 = \frac{1}{2}R_\gamma K_p \gamma L_1^2$$

$$P_2 = \frac{1}{2}R_\gamma K_p \gamma L_2^2$$

where:

K_p = passive earth pressure coefficient with $\delta = \phi$

R_γ = reduction factor for K_p which is a function of δ/ϕ

According to the earth pressure table presented by Caquot and Kerisel (1949), the value of K_p and R_γ can be estimated. In the case of assuming the actual failure surface, the mobilized friction angle is equal to the soil's friction angle, but if the simplified planes, ab and cd, are considered to be the failure surface, then the mobilized friction angle is an average value.

Fig. 7.12 illustrates the nature of distribution of the mobilized friction angle. Fig. 7.12 indicates that at points b, d the mobilized friction angle is equal to $\lambda\phi$ and the value of the mobilized friction angle at points a, c is equal to the soil's friction angle due to the fact that "a" and "c" are

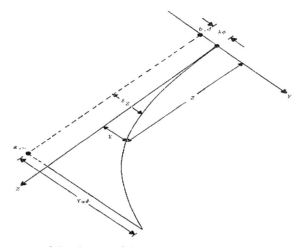

Figure 7.12 Nature of distribution of δ_z.

points that are placed on the actual failure surface. Hanna et al. also have carried out some laboratory tests that contributed to the presentation of the below equation for determination of the average value of the mobilized friction angle:

$$R_\gamma K_p \sin \delta = \frac{Q'_u - 0.5(L_1 + L_2)h \cdot \gamma \cdot \cos \psi}{0.5\gamma(L_1^2 + L_2^2)}$$

The preceding equation is the result of assimilation of some laboratory tests into the passive earth pressure coefficient table. In fact, the right hand side of the preceding equation is the result of some experimental tests. Eventually, both sides of the equation became parallel, which means that the average value of δ can be determined. Once the average value of δ was estimated, the variation of the locally mobilized angle of shearing resistance δ_Z could be calculated as follows:

$$P_1 + P_2 = \frac{1}{2}R_\gamma K_p \gamma(L_1^2 + L_2^2) = \gamma \left[\int_0^{L_1} K_{p(Z)}z \cdot dz + \int_0^{L_2} K_{p(Z)}z \cdot dz \right]$$

The results of several laboratory tests signify that the solution of the preceding equation can be determined if:

$$\lambda = \left(\frac{\psi}{90}\right)^3 + e^{-5 \tan \phi}$$

$$Y = \frac{A'Z}{1 - B'Z}$$

While:

$$A' = \frac{\lambda\phi}{\beta}$$

$$B' = \frac{1}{\beta}$$

It needs to be pointed out that β can be estimated according to boundary conditions. In fact, β could be determined along ab and cd by the following equations, respectively:

$$\beta = \frac{L_1}{1 - \lambda}$$

$$\beta = \frac{L_2}{1 - \lambda}$$

The procedure for determination of δ:
1. Estimate the value of the soil's friction angle
2. Determine the value of λ from the following equation:

$$\lambda = \left(\frac{\psi}{90}\right)^3 + e^{-5 \tan \phi}$$

3. Determine the variation of Y by the following equations:

$$Y = \frac{A'Z}{1 - B'Z}$$

$$A' = \frac{\lambda\phi}{\beta}$$

$$B' = \frac{1}{\beta}$$

$$\beta = \frac{L_1}{1 - \lambda}$$

$$\beta = \frac{L_2}{1 - \lambda}$$

4. Determine the value of δ_z by the below equation:

$$\delta_z = \lambda\phi + Y$$

5. Once the value of δ_z was determined, obtain the value of $K_{p(Z)}$ from the table presented by Caquot and Kerisel (1949)
6. Calculate the magnitude of $R_\gamma K_p$ from the following equation:

$$R_\gamma K_p = \frac{\int_0^{L_1} K_{p(Z)} z \cdot dz + \int_0^{L_2} K_{p(Z)} z \cdot dz}{0.5(L_1^2 + L_2^2)}$$

7. By using Caquot's and Kerisel's passive earth pressure table, the value of δ could be determined once the right hand side of the preceding equation was calculated.

Fig. 7.13 illustrates a sample of the aforementioned analysis. The figure shows the variation of δ/ϕ versus ψ for different soil friction angles. The preceding analysis can be simplified by the following assumption:

$$K_s \sin \phi = R_\gamma K_p \sin \delta$$

where:
K_s = punching uplift coefficient.
Fig. 7.14 illustrates the variation of K_s versus different values of ψ and ϕ.

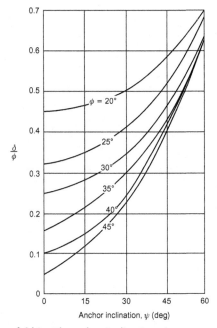

Figure 7.13 Variation of δ/ϕ with anchor inclination ψ.

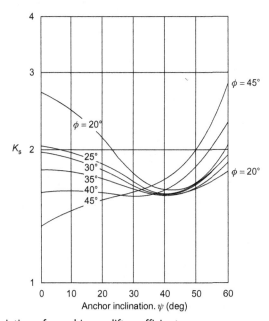

Figure 7.14 Variation of punching uplift coefficient.

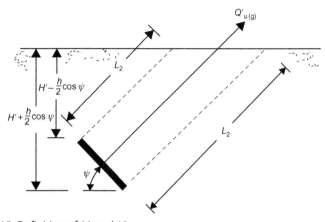

Figure 7.15 Definition of $L1$ and $L2$.

The net ultimate holding capacity per unit width of a plate anchor can be determined by the following equation which is the result of combination of the aforementioned equations.

$$Q'_u = P_p - P_a = cK_CH + \frac{1}{2}K_b\gamma H^2 + W \cos \psi$$

According to Fig. 7.15, $L1$ and $L2$ can be defined by the following equations:

$$L_1 = \frac{H' + \frac{h}{2}\sin \psi}{\cos \psi}$$

$$L_2 = \frac{H' - \frac{h}{2}\sin \psi}{\cos \psi}$$

So

$$Q'_u = \gamma K_s \frac{\sin \phi}{\cos^2_\psi}\left(H'^2 + \frac{h^2}{4}\sin^2 \psi\right) + \gamma H'h$$

Example 7.2

With the parameters for sand and anchor given in Example 7.1, determine Q_u' for $\psi = 0°, 45°$ and $60°$ using the theory of Hanna et al.

Solution

Given: $H' = 0.6$ m; $h = 0.2$; $\gamma = 18$ kN/m³; $\phi = 35°$ from Fig. 7.14 for $\phi = 35°$ the variation of K_s are as follows:

Anchor inclination, ψ (deg)	K_S
0	1.8
45	1.6
60	2.0

Using the below equation:

$$Q_u' = \gamma K_s \frac{\sin \phi}{\cos_\psi^2}\left(H'^2 + \frac{h^2}{4}\sin^2 \psi\right) + \gamma H' h$$

$\psi = 0$

$$Q_u' = (18)(1.8)\frac{\sin 35}{\cos^2 0}\left((0.6)^2 + \frac{(0.2)^2}{4}\sin^2 0\right) + (18)(0.6)(0.2) = 8.2544$$

$\psi = 45°$

$$Q_u' = (18)(1.6)\frac{\sin 35}{\cos^2 45}\left((0.6)^2 + \frac{(0.2)^2}{4}\sin^2 45\right) + (18)(0.6)(0.2) = 9.598$$

$\psi = 60°$

$$Q_u' = (18)(2)\frac{\sin 35}{\cos^2 60}\left((0.6)^2 + \frac{(0.2)^2}{4}\sin^2 60\right) + (18)(0.6)(0.2) = 22.116$$

7.3 RELATED THEORIES

Maiah et al. (1986) also conducted an investigation into determination of the ultimate holding capacity of shallow inclined anchors embedded in sand. It needs to be pointed out that the following equation was developed for shallow strip anchors but it is also applicable for rectangular anchors. The presented simple equation is as below:

$$Q_{u-\psi} = Q_{u-\psi=0°} + \left[Q_{u-\psi=90°} - Q_{u-\psi=0°}\right]\left(\frac{\psi}{90}\right)^2$$

$$Q_{u-\psi=90°} = \gamma h^2 B M_{\gamma q} S_f$$

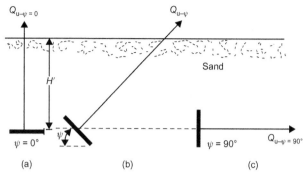

Figure 7.16 Definition of $Q_{u-\psi}$, $Q_{u-\psi=0}$ and $Q_{u-\psi=90}$ $Q_{u-\psi}$, $Q_{u-\psi=0}$ and $Q_{u-\psi=90}$.

While:

$Q_{u-\psi}$ = net ultimate holding capacity of anchor inclination of ψ with respect to the horizontal (Fig. 7.16a)

$Q_{u-\psi=0^\circ}$ = net ultimate uplift capacity of horizontal anchor ($\psi = 0$; Fig. 7.16b)

$Q_{u-\psi=90^\circ}$ = net ultimate holding capacity of vertical anchor ($\psi = 90^\circ$; Fig. 7.16c)

For estimation of $Q_{u-\psi=0^\circ}$ the following equations can be applied:

$$Q_{u-\psi=0^\circ} = \gamma H'^2 \left[2\left(1 + m\frac{H'}{h} \right)h + B - h \right] K_u \tan \phi$$
$$+ hBH'\gamma \text{ (Rectangular anchors)}$$

$$Q_{u-\psi=0^\circ} = \gamma H'^2 K_u \tan \phi + hH'\gamma \text{ (Strip anchor)}$$

where:

B = length of anchor plate

h = width of anchor plate

m = a coefficient for obtaining the shape factor

K_u = uplift coefficient (Fig. 7.17)

According to the surcharge theory presented by Neely et al. (1973), the following equation can be applied in order to estimate the value of $Q_{u-\psi=90^\circ}$:

$$Q_{u-\psi=90^\circ} = \gamma h^2 BM_{\gamma q} S_f$$

where

$M_{\gamma q}$ = according to Fig. 7.18, force coefficient

S_f = according to Fig. 7.19, shape factor

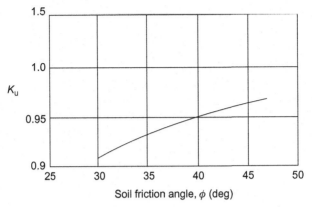

Figure 7.17 Meyerhof's uplift coefficient K_u for horizontal plate anchor.

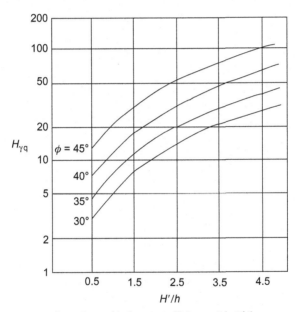

Figure 7.18 Variation of Neely et al.'s force coefficient with H'/h.

Example 7.3
Solve Example 7.1 using the aforementioned equations.

Solution
According to the following equation:

$$Q_{u-\psi=0°} = \gamma H'^2 K_u \tan \phi + h H' \gamma$$

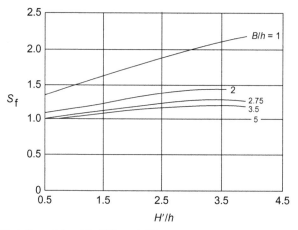

Figure 7.19 Variation of S_f with H'/h and B/h.

For

$$\phi = 35°, \ K_u = 0.93, \ h = 0.8 \text{ m}, \ H' = 0.6 \text{ m}, \ \gamma = 18 \text{ kN/m}^3$$

So

$$Q_{u-\psi=0°} = (18)(0.6)^2(0.93)\tan 35 + (0.8)(0.6)(18) = 3.692 + 8.64 = 12.332$$

Again from Fig. 7.19, for $H'/h = 2.5$ and $M_{\gamma q} \approx 20$ and $S_f = 1$ (strip anchor) so according to the following equation:

$$Q_{u-\psi=90°} = \gamma h^2 B M_{\gamma q} S_f$$

$$Q_{u-\psi=90°} = (18)(0.8)^2(1)(20)(1) = 54.4 \text{ kN/m}$$

Now, according to the following equation the value of $Q_{u-\psi}$ can be determined:

ψ (deg)	$Q'_{u-\psi}$ (kN/m)
0	17.68
45	26.86
60	34.00
75	43.13
90	54.4

$$Q_{u-\psi} = Q_{u-\psi=0°} + \left[Q_{u-\psi=90°} - Q_{u-\psi=0°}\right]\left(\frac{\psi}{90}\right)^2$$

Example 7.4

Compare the results of Examples 7.1–7.3.

Solution

The following conclusions can be made comparing the aforementioned equations:

1. Meyerhof's analysis of shallow inclined anchors and Miah et al.'s analysis gave quite close results.

$$Q'_u = \frac{1}{2}K_b\gamma\left(H' + \frac{h\sin\psi}{2}\right)^2 + \gamma h\left(H' + \frac{h\sin\psi}{2}\right)\cos^2\psi \quad \text{(Meyerhof's equation)}$$

$$Q_{u-\psi} = Q_{u-\psi=0^\circ} + \left[Q_{u-\psi=90^\circ} - Q_{u-\psi=0^\circ}\right]\left(\frac{\psi}{90}\right)^2 \quad \text{(Miah et al.'s equation)}$$

2. According to the theory of Hanna et al, the magnitude of $Q_{u-\psi}$ estimated higher than the actual value.

7.4 CONCLUSION

Due to the laboratory tests and observations discussed in this chapter, the following conclusions can be made:

1. The average value of the critical embedment ratio for circular anchors embedded horizontally $(\psi = 0^\circ)$ and vertically $(\psi = 90^\circ)$ in sand are determined as below:

$$\psi = 0^\circ: \quad \text{Loose sand} \quad 4$$
$$\text{Dense sand} \quad 8$$
$$\psi = 90^\circ: \quad \text{Loose sand} \quad 4$$
$$\text{Dense sand} \quad 6$$

2. Also, the critical embedment ratio for strip anchors embedded in sand are as follows:

$$\psi = 0^\circ: \quad \text{Loose sand} \quad 6$$
$$\text{Dense sand} \quad 11-12$$
$$\psi = 90^\circ: \quad \text{Loose sand} \quad 4$$
$$\text{Dense sand} \quad 8$$

3. The magnitude of $Q_{u-\psi}$ can be determined by Meyerhof's equation or the following equations. However, for shallow rectangular anchors only the following equations can be applied:

$$Q_{u-\psi} = Q_{u-\psi=0^\circ} + \left[Q_{u-\psi=90^\circ} - Q_{u-\psi=0^\circ}\right]\left(\frac{\psi}{90}\right)^2$$

$$Q_{u-\psi=0^\circ} = \gamma H'^2 \left[2\left(1 + m\frac{H'}{h}\right)h + B - h\right]K_u \tan\phi + hBH'\gamma$$

(Rectangular anchor)

$$Q_{u-\psi=0^\circ} = \gamma H'^2 K_u \tan\phi + hH'\gamma \quad \text{(Strip anchor)}$$

4. For deep anchors, the following equation is applicable. It is worth noting that the following equation can be applied for strip, square and rectangular anchors. However, the values of $F'^*_{q-\psi=0^\circ} = Q'_{u-\psi=0^\circ}/\gamma hH'$ and $F'^*_{q-\psi=90^\circ} = Q'_{u-\psi=90^\circ}/\gamma hH'$ for strip anchors should be determined from Fig. 7.8. Also, the values of $F'^*_{q-\psi=90^\circ}$ and $F'^*_{q-\psi=90^\circ}$ can determined from Fig. 7.9. It needs to be pointed out that for rectangular anchors, the value of $F'^*_{q-\psi}$ can be determined by an interpolation between the breakout factor for $\psi = 90^\circ$ and $\psi = 0^\circ$. The following equation can be applied for estimation of the value of the ultimate load for a known magnitude of breakout factor:

$$Q'_{u-\psi} = F'^*_{q-\psi}\gamma h^2 H' \quad \text{(Square anchor)}$$

$$Q'_{u-\psi} = F'^*_{q-\psi}\gamma hH' \quad \text{(Strip anchor)}$$

$$Q'_{u-\psi} = F'_{q-\psi}\gamma hBH' \quad \text{(Rectangular anchor)}$$

5. As the inclination angle of an anchor increases, the anchor's displacement increases as well. According to an analysis based on laboratory tests, Das et al. investigated that the value of Δ_u/h increases as the value of H'/h increases. Also the proper value of Δ_u/h have been estimated for shallow anchors as below:

Anchor type	Δ/h at $\Psi = 0$ (%)	Δ/h at $\Psi = 90$(%)
Strip	6−8	10−25
Square	8−10	15−30

6. It is recommended that a safety factor of 3 be incorporated in the results of the ultimate holding capacity.
7. It is clear that further studies needs to be carried out in order to investigate the effects of center to center spacing between inclined anchors.

REFERENCES

Baker, W.H., Konder, R.L., 1996. Pullout Load Capacity of Circular Earth Anchor Buried in Sand. Highway Research Record No. 108. National Academy of Sciences, Washington, D.C., pp. 1—10.

Balla, A., 1961. The resistance of breakout of mushroom foundations for pylons. Proc, V Intl. Conf. Soil Mech. Found. Eng. Paris, 1. pp. 569—576.

Brinch Hansen, J., 1961. The ultimate resistance of rigid piles against transverse forces. Bull. No. 12, Danish Geotech. Inst. Copenhagen, Denmark.

Caquot, A., Kerisel, L., 1949. Traite de Mechnique des Sols. Gauthier-Villars, Paris.

Das, B.M., 1990. Earth Anchors. Elsevier, Amsterdam.

Hanna, A.M., Das, B.M., Foriero, A. 1988. Behavior of shallow inclined plate anchors in sand. In Spec. Topics in Found. Geotech. Spec. Tech. Publ. No. 16. ASCE. pp. 54—72.

Harvey, R.C., Burley, E., 1973. Behavior of shallow inclined anchorages in cohesion-less sand. Ground Eng. 6 (5), 48—55.

Kananyan, A.S., 1966. Experimental investigation of the stability of bases of anchor foundations. Osnovanlya Fundamenty i mekhanik Gruntov 4 (6), 387—392.

Maiah, A.A., Das, B.M., Picornell, M. 1986. Ultimate resistance of shallow inclined strip anchor plate in sand. Proc, Southeastern Conf. on Theoretical and Applied Mech. Columbia, SC, USA 2. pp. 503—509.

Meyerhof, G.G., 1973. Uplift resistance of inclined anchors and piles. Proc, VIII Intl. Conf. Soil Mech. Found. Eng. Moscow. USSR 2 (1),167—172.

Meyerhof, G.G., Adams, J.I., 1968. The ultimate uplift capacity of foundations. Can. Geotech. J. 5 (4), 225—244.

Neely, W.J., Stuart, J.G., Graham, J., 1973. Failure load of vertical anchor plates in sand. J. Geotech. Eng. Div. ASCE. 99 (9), 669—685.

Rankine, W., 1857. On the stability of loose earth. Philos. Trans. R. Soc. Lond. 147.

Sokolovskii, V.V., 1965. Statics of Granular Media. Pergamon Press, London.

CHAPTER 8

Inclined Anchor Plates in Cohesive Soil

8.1 INTRODUCTION

As we have seen in chapter "Inclined Anchor Plates in Cohesionless Soil," plate anchors are a subcategory of anchors that are applied in a structure's foundation in order to resist the outwardly-directed load imposed on it. Anchors can be located in the soil in three different directions: horizontal, vertical, and inclined. Both vertical and inclined anchors can resist the pullout force and overturning movements, although inclined anchors are more flexible for bearing overturning situations.

Inclined anchors are used for foundations where plate anchors should be located in an inclination to the horizontal. Inclined anchors are applicable in many different types of structures, such as electric power transmission towers and retaining walls supporting deep excavation. It has also been mentioned in previous chapters that an anchor's load-bearing capacity stems from the passive resistance of soil. Researchers such as Ghaly and Hanna (1994) have conducted a study into investigation of performance of axially-loaded inclined anchors. According to Ghaly and Hanna's study, the behavior of anchors with the inclination angle of 15°, 30°, and 45° with respect to the vertical was compared with the performance of vertical anchors in the same condition. Based on these experimental tests, Ghaly and Hanna presented a model for ultimate holding capacity of inclined anchors. Other researchers such as Robinson and Taylor (1969), Vesic (1971), Meyerhof (1973), and Larnach and McMullan (1974), have carried out investigations of the performance of inclined anchors, although as highlighted earlier in this book, much of this research was devoted to anchors embedded in sand. Robinson and Taylor (1969) mostly focused on multi-helix screw anchors embedded in soil with inclination angles of 10°, 23°, and 30° with respect to the horizontal. The results showed that as the inclination angle increases, the ultimate pullout capacity of the anchor increases as well. Vesic (1971) also investigated objects embedded at the bottom of the ocean. According to

Design and Construction of Soil Anchor Plates.
DOI: http://dx.doi.org/10.1016/B978-0-12-420115-6.00008-4

Vesic's study, objects oriented at an angle of $0-45°$ possess almost double the breakout force. Also, Vesic's analysis revealed that the failure pattern of the soil with objects embedded in it, may vary with the depth of embedment. Meyerhof (1973) also concentrated on the behavior of inclined anchors and compared them with vertical anchors. Meyerhof's studies proposed that the pullout capacity of inclined anchors is more than vertical anchors that are axially loaded. Larnach and McMullan (1974) also carried out a study that was aimed at comparing the performance of inclined groups of plate anchors with single anchors. The indicated that the maximum peak load of individual inclined anchors occur at an inclination angle of $20°$ while the maximum peak load of group anchors take place at an inclination angle of $35°$.

The main aim of all this research was to determine the ultimate holding capacity of inclined anchors, although a high proportion of these studies were concentrated on anchors embedded in sand. In this chapter, we show how to determine the ultimate holding capacity of inclined anchors embedded in clay.

8.2 EARLY THEORIES ON INCLINED ANCHOR PLATES

Chapter "Inclined Anchor Plates in Cohesionless Soil" looked at determining the ultimate holding capacity of horizontal anchors and vertical anchors that were embedded in clay. Unlike the preceding chapters, the existing studies that have been conducted for determining the ultimate holding capacity of inclined anchors are really limited. One of the researchers who has thoroughly investigated inclined anchors is Das (1985) whose analysis is going to be presented here. The analysis carried out by Das was mainly founded on laboratory tests conducted on square anchors embedded in clay soils that are saturated or almost saturated. Based upon trial-and-error and experimental tests, Das presented the following equation for determinating the ultimate holding capacity of inclined rectangular anchors:

$$Q_u = Ac_u F'_c + W \cos \psi$$

where:

A = plate anchor's area which is B^*h

B = width of the plate anchor

c_u = undrained cohesion of the clay soil whose friction angle is equal to Zero

F'_c = average breakout factor

W = weight of soil located immediately above anchor

ψ = inclination angle of the anchor with respect to horizontal

The weight of soil located immediately above anchor could be determined as follows:

$$W = A\gamma H' \cos \psi$$

According to Fig. 8.1, H' is the average depth of embedment.

By assimilating the below equation:

$$W = A\gamma H' \cos \psi$$

into the aforementioned equation presented for determining the ultimate holding capacity of inclined anchor, the following equation can be obtained:

$$F'_c = \frac{\frac{Q_u}{A} - \gamma H' \cos^2 \psi}{c_u}$$

Also:

$$F'_c = \frac{\frac{Q_u}{h^2} - \gamma H' \cos^2 \psi}{c_u} \quad \text{(For square anchors)}$$

$$F'_c = \frac{\frac{Q_u}{hB} - \gamma H' \cos^2 \psi}{c_u} \quad \text{(For rectangular anchors)}$$

$$F'_c = \frac{\frac{Q_u}{h} - \gamma H' \cos^2 \psi}{c_u} \quad \text{(For strip anchors)}$$

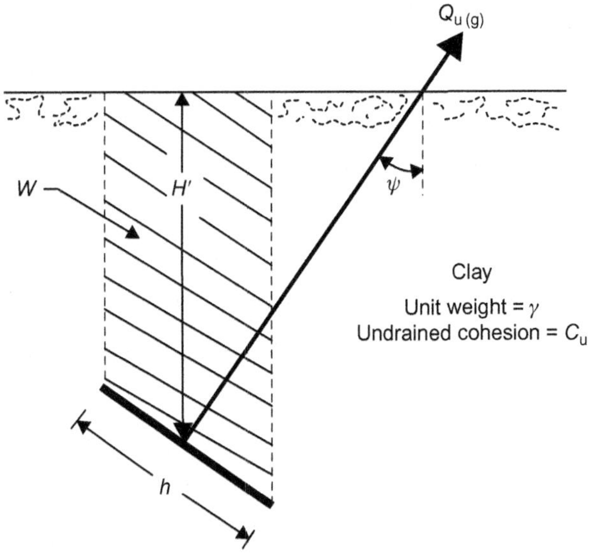

Figure 8.1 Inclined plate anchor in clay.

The average breakout factor can be determined according to the following equation:

$$F'_{c-\psi} = F'_{c-\psi=0°} + (F'_{c-\psi=90°} - F'_{c-\psi=0°})\left(\frac{\psi°}{90}\right)^2$$

where:

$F'_{c-\psi}$ = average breakout factor of anchor oriented at the angle of ψ with respect to the horizontal

$F'_{c-\psi=0°}$ = average breakout factor of horizontal anchor

$F'_{c-\psi=90°}$ = average breakout factor of vertical anchor

It needs to be pointed out that there is a direct relationship between the values of $F'_{c-\psi}$ and the average embedment ratio of the inclined anchor. In fact, as the value of the embedment ratio H'/h increases, the magnitude of $F'_{c-\psi}$ increases as well up to a maximum value of $F'^*_{c-\psi}$ which occurs in critical embedment ratio of $(H'/h)_{cr}$. The value of $F'_{c-\psi}$ remains constant for embedment ratios above the critical embedment ratio (see Fig. 8.2).

Das (1980) and Das et al. (1985) have carried out many laboratory tests in order to determine the values of $F'_{c-\psi=0°}$ and $F'_{q-\psi=90°}$. Based on the aforementioned analyses, the following procedure is presented for estimation of $F'_{c-\psi=0°}$ and $F'_{q-\psi=90°}$.

8.2.1 Determination of $F'_{c-\psi=0°}$

1. First, the critical embedment ratio $(H'/h)_{cr-R}$ for rectangular anchor should be calculated by the following equation:

$$\left(\frac{H'}{h}\right)_{cr-R} = \left(\frac{H'}{h}\right)_{cr-S}\left[0.73 + 0.27\left(\frac{B}{h}\right)\right] \le 1.55\left(\frac{H'}{h}\right)_{cr-S}$$

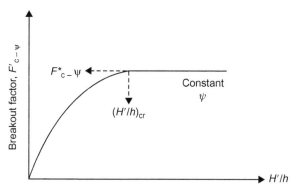

Figure 8.2 Variation of $F'_{c-\psi}$ with H'/h.

And

$$\left(\frac{H'}{h}\right)_{cr-S} = 0.107\ c_u + 2.5 \leq 7$$

2. Second, if the actual embedment ratio of the rectangular anchor which was calculated in previous step was greater than $(H'/h)_{cr}$ then the anchor is a deep anchor. The following equation can be applied for deep anchors:

$$F'_{q-\psi=0^\circ} = F^{*'}_{q-\psi=90^\circ} = 7.56 + 1.44\left(\frac{h}{B}\right)$$

In the case of actual embedment ratio less than the critical embedment ratio, the anchor is a shallow anchor. The preceding equation can be presented as follows for shallow anchors (see also Fig. 8.3):

$$F'_{q-\psi=0^\circ} = \left[7.56 + 1.44\left(\frac{h}{B}\right)\right]\beta$$

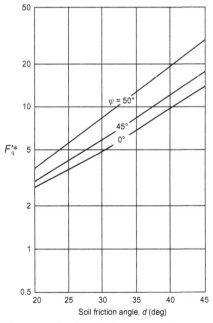

Figure 8.3 Variation of Meyerhof's F'^{*}_{q} with friction angle.

While according to Fig. 8.4:

$$\beta = f \left[\frac{\dfrac{H'}{h}}{\left(\dfrac{H'}{h} \right)_{cr}} \right]$$

8.2.2 Determination of $F'_{c-\psi=90°}$

In the first step the value of the critical embedment ratio for rectangular anchor should be calculated:

$$\frac{\left(\dfrac{H'}{h} \right)_{cr-R} + 0.5}{\left(\dfrac{H'}{h} \right)_{cr-s} + 0.5} = \left[0.9 + 0.1 \left(\frac{B}{h} \right) \right] \le 1.31$$

where:

$\left(\frac{H'}{h} \right)_{cr-s}$ = critical embedment ratio of a square anchor measuring $h \times h$

Note that the critical embedment ratio for square anchors can be determined as follows:

$$\left(\frac{H'}{h} \right)_{cr-s} = 4.2 + 0.0606\, c_u \le 6.5$$

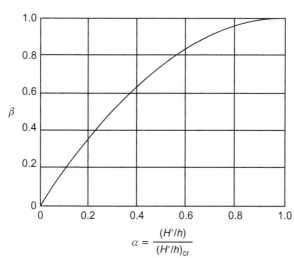

Figure 8.4 Variation of β with $(H'/h)/(H'/h)_{cr}$ for $\psi = 0°$.

Once the critical embedment ratio is calculated, compare the actual embedment ratio with $(H'/h)_{cr}$. If (H'/h) is greater than $(H'/h)_{cr}$, then it is a deep anchor.

$$F'_{c-\psi=90°} = (9)\left[0.825 + 0.175\left(\frac{h}{B}\right)\right]$$

If the actual embedment ratio is less than $(H'/h)_{cr}$, then it is a shallow anchor.

$$\frac{F'_{c-\psi=90°}}{F^{*'}_{c-\psi=90°}} = \frac{n'}{0.41 + 0.59\,n'}$$

While:

$$n' = \frac{\left(\dfrac{H'}{h}\right) + 0.5}{\left(\dfrac{H'}{h}\right)_{cr} + 0.5}$$

According to the following equation, the magnitude of $F'_{c-\psi}$ can be calculated:

$$F'_{c-\psi} = F'_{c-\psi=0°} + (F'_{c-\psi=90°} - F'_{c-\psi=0°})\left(\frac{\psi°}{90}\right)^2$$

And the ultimate holding capacity, assuming that $F'_{c-\psi}$ is known can be calculated by the following equations:

$$F'_c = \frac{\dfrac{Q_u}{h^2} - \gamma H' \cos^2\psi}{c_u} \qquad \text{(For square anchors)}$$

$$F'_c = \frac{\dfrac{Q_u}{hB} - \gamma H' \cos^2\psi}{c_u} \qquad \text{(For rectangular anchors)}$$

$$F'_c = \frac{\dfrac{Q_u}{h} - \gamma H' \cos^2\psi}{c_u} \qquad \text{(For strip anchors)}$$

Example 8.1

For an anchor embedded in a saturated clay, with the following information, $h = 0.2$ m, $H' = 0.6$, $B = 0.4$ m, $\psi = 30°$, for the clay: $c_u = 28$ kN/m, $\gamma = 17$ kN/m^3, calculate the net ultimate holding capacity.

Solution

Calculation of $F'_{c-\psi=0°}$:

The critical value of the embedment ratio for square anchor should be calculated by the following equation:

$$\left(\frac{H'}{h}\right)_{cr-S} = 0.107\, c_u + 2.5 \leq 7$$

So

$$\left(\frac{H'}{h}\right)_{cr-S} = 0.107\, c_u + 2.5 = (0.107)(28) + 2.5 \approx 5.5$$

The critical embedment ratio for rectangular anchor can be estimated by the following equation:

$$\left(\frac{H'}{h}\right)_{cr-R} = \left(\frac{H'}{h}\right)_{cr-S}\left[0.73 + 0.27\left(\frac{B}{h}\right)\right] \leq 1.55\left(\frac{H'}{h}\right)_{cr-S}$$

So

$$\left(\frac{H'}{h}\right)_{cr-R} = \left(\frac{H'}{h}\right)_{cr-S}\left[0.73 + 0.27\left(\frac{B}{h}\right)\right] = 5.5\left[0.73 + 0.27\left(\frac{0.4}{0.2}\right)\right] \approx 6.99$$

Due to the fact that 6.99 is less than $(1.55) \times (5.55) = 8.525$, the critical embedment ratio for rectangular anchors, $\left(\frac{H'}{h}\right)_{cr-R} = 6.99$, should be used in equations. Also, the embedment ratio $H'/h = 1.2/0.4 = 3$ so, it is a shallow anchor and the breakout factor for inclination angle of zero can be determined by the following equation:

$$F'_{q-\psi=0°} = \left[7.56 + 1.44\left(\frac{h}{B}\right)\right]\beta$$

According to Fig. 8.4, the ratio of $\dfrac{\left(\frac{H'}{h}\right)}{\left(\frac{H'}{h}\right)_{cr}} = 3/6.99, \beta = 0.69$

So

$$F'_{q-\psi=0°} = \left[7.56 + 1.44\left(\frac{0.2}{0.4}\right)\right](0.69) = 5.71$$

Calculation of $F'_{q-\psi=90°}$:

First of all the critical embedment ratio for square anchor should be calculated by the following equation:

$$\left(\frac{H'}{h}\right)_{cr-s} = 4.2 + 0.0606\, c_u \leq 6.5$$

So

$$\left(\frac{H'}{h}\right)_{cr-s} = 4.2 + (0.0606)(28) = 5.9$$

The critical embedment ratio for rectangular anchor can be calculated by the following equation for a given $\left(\frac{H'}{h}\right)_{cr-s}$:

$$\frac{\left(\frac{H'}{h}\right)_{cr-R} + 0.5}{\left(\frac{H'}{h}\right)_{cr-s} + 0.5} = \left[0.9 + 0.1\left(\frac{B}{h}\right)\right] \leq 1.31$$

$$\frac{\left(\frac{H'}{h}\right)_{cr-R} + 0.5}{\left(\frac{H'}{h}\right)_{cr-s} + 0.5} = \left[0.9 + 0.1\left(\frac{0.4}{0.2}\right)\right] = 0.9 + (0.1)(2) = 1.1$$

$$\left(\frac{H'}{h}\right)_{cr-R} + 0.5 = 1.1\left[\left(\frac{H'}{h}\right)_{cr-s} + 0.5\right] = 1.1(5.9 + 0.5) = 7.04$$

$$\left(\frac{H'}{h}\right)_{cr-R} = 6.54$$

As the embedment ratio is less than the critical embedment ratio, the anchor is shallow. According to the following equations $F'_{c-\psi=90°}$ can be calculated.

$$\frac{F'_{c-\psi=90°}}{F^{*'}_{c-\psi=90°}} = \frac{n'}{0.41 + 0.59\,n'}$$

$$\frac{n'}{0.41 + 0.59\,n'} = \frac{3 + 0.5}{6.54 + 0.5} = 0.497$$

$$F^{*'}_{c-\psi=90°} = (9)\left[0.825 + 0.175\left(\frac{h}{B}\right)\right] = 9\left[0.825 + 0.175\left(\frac{0.2}{0.4}\right)\right]$$

The value of $F'_{c-\psi=90°}$ can be calculated according to the aforementioned equation:

$$F'_{c-\psi=90°} = \frac{n' \cdot F^{*'}_{c-\psi=90°}}{0.41 + 0.59\,n'}$$

So

$$F'_{c-\psi=90°} = \frac{(8.21).(0.497)}{0.41 + (0.59)(0.497)} = 5.8$$

According to the following equation, the ultimate holding capacity can be calculated:

$$Q_u = Bh(c_u F'_{c-\psi} + \gamma H' \cos^2 \phi)$$

$$Q_u = (0.4)(0.2)((28)(5.72) + (17)(1.2)\cos^2 30) = 14.215$$

REFERENCES

Das, B.M., 1980. A procedure for estimation of ultimate uplift capacity of foundations in clay. Soils Found. 20 (1), 72–82.

Das, B.M., 1985. Resistance of shallow inclined anchors in clay. In: Clemence, S.P. (Ed.), Uplift Behavior of Anchor Found in Soils. ASCE, pp. 86–101.

Das, B.M., 1990. Earth Anchors. Elsevier, Amsterdam.

Das, B.M., Tarquin, A.J., Moreno, R., 1985. Model tests for pullout resistance of vertical anchors clay. Civ. Eng. Pract. Design Eng. 4 (2), 191–209, Pergamon Press.

Ghaly, A.M., Hanna, A.M., 1994. Ultimate pullout resistance of vertical anchors. Can. Geotech. J. Ottawa: Canada. 31 (5), 661–672.

Larnach, W.J., McMullan, D.J. 1974. Behaviour of inclined groups of plate anchors in dry sand. Proc. Conj. Inst. Civ. Eng. London: England. pp. 153–156.

Meyerhof, G.G., 1973. Uplift resistance of inclined anchors and piles. Proc. VIII Intl. Conf. Soil Mech. Found. Eng. Moscow 2 (1), 167–172, USSR.

Robinson, K.E., Taylor, H., 1969. Selection and performance of anchors for guyed transmission towers. Can. Geotech. J. Ottawa: Canada 6 (2), 119–137.

Vesic, A.S., 1971. Breakout resistance of objects embedded in ocean bottom. J. Soil Mech. Found. Div. ASCE. 97 (9), 1183–1205.

CHAPTER 9

Anchor Plates in Multilayer Soil

9.1 INTRODUCTION

Most soils are made up of multiple layers. Whilst soil anchor plates can of course be used in multilayer soils there has only been minimal research dedicated to this area (Bouazza and Finlay, 1990; Krishna, 2000; Stewart, 1985; Manjunath, 1998; Niroumand et al., 2010, 2011). We have seen in previous chapters that the use of deep anchor plates in clay results in the ultimate uplift capacity remaining constant after an embedment ratio of more than 4.5, and sand soils increased the ultimate uplift capacity with increasing the embedment ratio. Combining these to have a multilayer soil is a good solution for increasing the ultimate uplift capacity of clay layers (by placing a sand layer over clay layers). This chapter discusses this concept in further detail.

9.2 EARLY THEORIES ON ANCHOR PLATES IN MULTILAYER SOIL

During the past few years, a great number of experimental model and numerical analysis results on the uplift resistance of anchor plate embedded in homogeneous soil has been published. However, a review of related literature shows that very little research has analyzed the performance of anchor plates in layered soils: a problem, which is often encountered by the professional engineers in the field. Fig. 9.1 shows the two-layered cohesionless soils for pullout loading of a layer of loose sand overlying dense sand.

9.2.1 Stewart's Method

Stewart (1985) evaluated the effectiveness of placement of sand layers on clay layers for increasing the uplift capacity of soil anchor plates located in clay layer. Fig. 9.2 illustrates the general view of experiments set up by Stewart. He found that he could increase the uplift capacity of soil anchor plates in clay layer by having a sand layer overlay the clay layer as illustrated in Fig. 9.3. Some results shown by Stewart's method were: (1) the sand layer overlaying the clay layer increased the uplift capacity of circular

Design and Construction of Soil Anchor Plates.
DOI: http://dx.doi.org/10.1016/B978-0-12-420115-6.00009-6

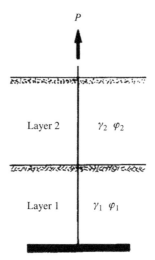

Figure 9.1 Anchor plate under pullout load in two-layered cohesionless soils.

anchor plate located in the clay layer; (2) a dense sand layer showed a better uplift capacity rather than loose sand overlaying the clay layer; (3) the uplift capacity increased for a circular anchor plate located in clay layer when embedment ratio was not more than 4.5, because after this amount has no effect on the clay layer and its uplift capacity.

9.2.2 Manjunath's Method

Manjunath (1998) evaluated cohesion (F_c), surcharge (F_q) and unit weight (F_γ) for uplift capacity of a shallow horizontal strip anchor plate in two layered frictional- cohesive soils as illustrated in Fig. 9.4. He assumed the failure surface in two layers would approximate the standard log-spiral. Manjunath's method suggested the below function as:

$$P_{u-net} = (d_t c_t + d_b c_b)F_c + qF_q + 0.5B(d_t \gamma_t + d_b \gamma_b)F_\gamma$$

and the average ultimate uplift capacity (P_u) as

$$P_u = P_{u-net} + (d_t \gamma_t + d_b \gamma_b)$$

9.2.3 Bouazza and Finlay's Method

Bouazza and Finlay (1990) reported the behavior of an anchor plate buried in a two-layered cohesionless soil. The testing program consisted of a 37.5 mm diameter circular anchor plate buried in dense sandy soil

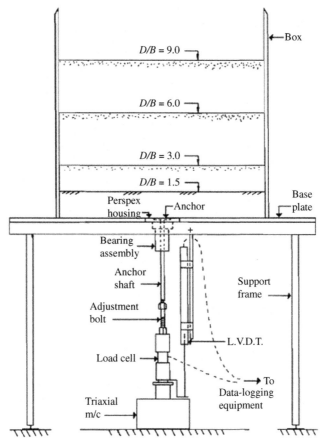

Figure 9.2 Uplift setup by Stewart (1985).

overlaid by loose or medium-dense sandy soil as shown in Fig. 9.5. The pullout tests were carried out on an anchor plate embedded at a depth (D) in a combination of layers of sand. The thickness of each layer was increased to a certain proportion of the anchor diameter and it was increased from 1 to 4 times the anchor diameter. It was reported that for upper layer thickness ratio of less than 1 and for a given embedment ratio (D/B), there was no difference between the pulling an anchor plate from a dense-medium bed or a dense-loose bed. For a given D/B ratio and the upper layer thickness ratio of 1−4 a dense-medium bed gives a greater pullout than a dense-loose bed as shown in Fig. 9.6. It is observed that the ultimate uplift capacity is dependent on the relative strength of the two layers, the depth ratio of embedment, and the upper layer thickness ratio.

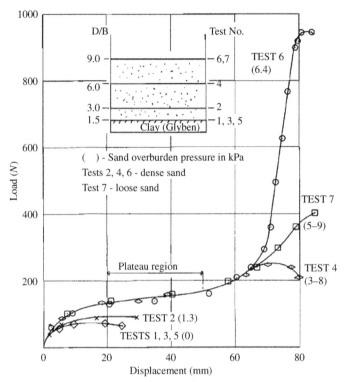

Figure 9.3 Net load versus anchor plate displacement in a clay layer overlaying a dense sand layer.

9.2.4 Krishna's Method

Krishna (2000) investigated the behavior of large size anchor plates in two-layered sand using an explicit two-dimensional finite difference program (FLAC 2D). Soil is assumed to be a Mohr−Coulomb strain softening/hardening material. The geotechnical properties of backfill of anchor foundations are very sensitive to construction and compaction methods. There is no satisfactory method to analyze the behavior of anchor plates in such nonhomogeneous cohesionless soil conditions. The two-layered soil for this analysis consisted of two cases: (1) a layer of loose sand overlaid by a dense sand layer and (2) a layer of dense sand overlaid by a loose sand layer, as shown in Fig. 9.7. For the analysis he chose sections of Chattahoochee River sand both in dense and loose conditions (Vesic and Clough, 1968). In the analyses, the width of the anchor plate (B) was 1 m and the embedment ratio was varied from 2 to 8. The upper layer

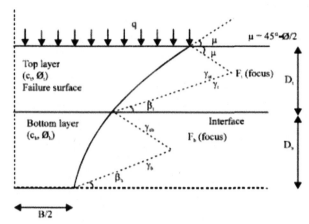

Figure 9.4 Failure mechanism in two layered soils by Manjunath (1998). γ_{ot}, Initial length of radial line of log-spiral failure surface in the top layer; γ_{ob}, Initial length of radial line of log-spiral failure surface in the bottom layer; γ_t, Final length of radial line of log-spiral failure surface in the top layer; γ_b, Final length of radial line of log-spiral failure surface in the bottom layer; β_t, angle between the final radial line of log-spiral failure surface in the top layer and the horizontal at the interface.

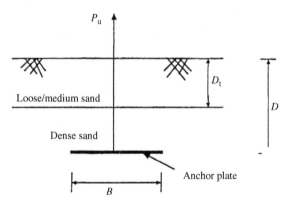

Figure 9.5 Experimental investigations layered soil system used by Bouazza and Finlay (1990).

thickness (D) is varied from minimum of B to maximum of ($D+2B$). The material properties of the anchor plate were kept constant. It was assumed that the plate was sufficiently stiff as not to affect the pullout response. Fig. 9.8 shows the variation of width for different D/B ratios. The ultimate pullout capacity changes, with an increase where the bottom layer is dense sand and top layer is loose sand. Fig. 9.9 shows displacement vectors and plastic regions at failure in layered cohesionless soils.

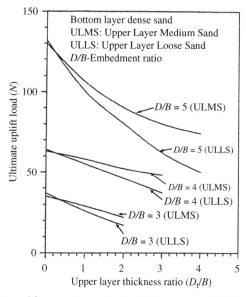

Figure 9.6 Ultimate uplift capacity against ratio (Bouazza and Finlay, 1990).

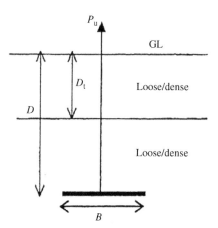

Figure 9.7 Anchor plate analyzed by Krishna (2000).

9.3 LIMITATIONS OF THE EXISTING STUDIES

Most research in this area has focused on two-layer soils, however in practice it is more likely that multiple layers are found onsite. More work needs to be done in analyzing the design and use of soil anchor plates and/or improving clay layers to resolve this problem.

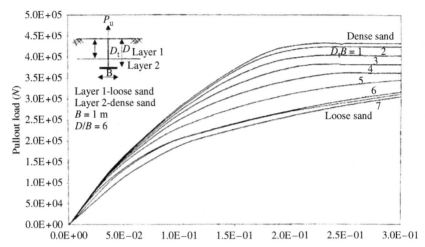

Figure 9.8 Ultimate uplift capacity against ratio by Krishna (2000).

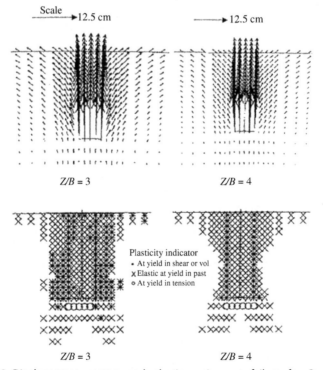

Figure 9.9 Displacement vectors and plastic regions at failure for $B = 1$ m and embedment ratio, $D/B = 4$ in layered soils in FLAC 2D by Krishna (2000).

9.4 CONCLUSION

There is limited research and results regarding multilayer soils and the use of soil anchor plates. Two-layer soil gives a greater embedment ratio and uplift capacity of soil anchor plates. Having a sand layer overlaying a clay layer can increase the uplift capacity of an anchor plate located in the clay layer and give a net ultimate uplift capacity as calculated by:

$$Q_u = Q_{u-\text{clay}} + Q_{u-\text{sand}}$$

REFERENCES

Bouazza, A., Finlay, T.W., 1990. Uplift capacity of plate anchors buried in two layered sand. Geotechnique 40, 293–297.

Das, B.M., 1990. Earth Anchors. Elsevier, Amsterdam.

Krishna, Y.S.R., 2000. Numerical analysis of large size horizontal strip anchors. Indian Institute of Science, Ph.D. Thesis.

Manjunath, K., 1998. Uplift capacity of horizontal strip and circular anchors in homogeneous and layered soils. Department of Civil Eng., Indian Institute of Science, Bangalore, Ph.D. Thesis.

Niroumand, H., Kassim, K.A., Nazir, R., 2010. Anchor plates in two-layered cohesion less soils. Am. J. Appl. Sci. 7 (10), 1396–1399.

Niroumand, H., Kassim, K.A., Nazir, R., 2011. Uplift capacity of anchor plates in two-layered cohesive-frictional soils. J. Appl. Sci. 11, 589–591.

Stewart, W., 1985. Uplift capacity of circular plate anchors in layered soil. Can. Geotech. J. 22, 589–592.

Vesic, A.S., Clough, G.W., 1968. Behavior of granular materials under high stresses. ASCE J. Soil Mech. Found. Div. 94, 661–688.

INDEX

Note: Page numbers followed by "*f*" refer to figures.

Printed in the United States
By Bookmasters